Third Street Books

320 NE Third Street
McMinnville, OR 97128
(503) 472-7786

Customer name: Karl Geislinger

Transaction #:0000647004
Station:ST2, 2 Clerk:STATION2
Thursday, June 5 2025   5:04 PM

SALES:
1@ 27.00 9781837260539	27.00
Nature's Genius	
SUBTOTAL	27.00
GRAND TOTAL	27.00
TENDER:	
Mc/visa XXXXXXXXXXXX8404	27.00
TOTAL TENDER	27.00
CHANGE	$0.00

TYPE: Credit
CARD: VISA
ENTRY: MANUAL
TRAN ID: 710858326
APPROVAL CODE: 981711

Current Book Club total: 104.97 Dollar(s)

Thank you! Unread books in new condition can be returned within 30 days of purchase with receipt. Check our website for news: thirdstreetbooks.com

Please visit us at www.thirdstreetbooks.com

00647004

CUSTOMER COPY

Third Street Books

220 NE Third Street
McMinnville OR 97128
(503) 472-7786

Cashier name: Kali Gestinger
Transaction #: 000064700/2
Station: 2, 2 CLERKSTATION/
Thursday, June 5, 2025   5:04 PM

SALES:
(x) 2.1.DD.9 8187260959   25.00
Beatrice's Ledger
SUBTOTAL   25.00

GRAND TOTAL   25.00

TENDER
Mc/Visa XXXXXXXXX8404   25.00
TOTAL TENDER   25.00

CHANGE   0.00

TYPE: CREDIT
CARD: VISA
ENTRY: MANUAL
TRAN ID: 710586256
APPROVAL CODE: 96V1Z1

Current book club total: 104.97 Dollar(s)

Thank you! Unread books in new
condition can be returned with a
30 days of purchase with receipt.
Check our website for news.
thirdstreetbooks.com

Please visit us at www.thirdstreetbooks.com

006470004

CUSTOMER COPY

Nature's Genius

Nature's Genius

Evolution's Lessons for a Changing Planet

David Farrier

CANONGATE

First published in Great Britain, the USA and Canada in 2025 by Canongate Books Ltd, 14 High Street, Edinburgh EH1 1TE

Distributed in the USA by Publishers Group West and in Canada by Publishers Group Canada

canongate.co.uk

1

Copyright © David Farrier, 2025

The right of David Farrier to be identified as the author of this work has been asserted by him in accordance with the Copyright, Designs and Patents Act 1988

No part of this book may be used or reproduced in any manner for the purpose of training artificial intelligence technologies or systems. This work is reserved from text and data mining (Article 4(3) Directive (EU) 2019/790).

Extract from Franz Kafka's notebooks on p. 247 translated by the author.

For Picture Credits please see p. 275

British Library Cataloguing-in-Publication Data

A catalogue record for this book is available on request from the British Library

ISBN 978 1 83726 053 9
Export ISBN 978 1 83726 054 6

Typeset in Garamond MT Std by Palimpsest Book Production Limited, Falkirk, Stirlingshire

Printed and bound by CPI Group (UK) Ltd, Croydon CR0 4YY

The manufacturer's authorised representative in the EU for product safety is Authorised Rep Compliance Ltd, 71 Lower Baggot Street, Dublin D02 P593 Ireland (arccompliance.com)

For all who choose hope over darkness

Indeed, what Reason may not go to Schoole to the wisdome of Bees, Ants and Spiders?
Thomas Browne

CONTENTS

Introduction   1

1. Optimum Dog   13
   How domestication proves change is possible

2. The Living City   41
   How urban evolution can teach us to build sustainable cities

3. One Touch Makes the Whole World Kin   75
   How nature can help us fix our waste problem

4. The Kinship of Languages   107
   How animal song can teach us to listen to other species

5. Strange Minds   143
   How other kinds of intelligence can help us remake our economies

6. Wild Clocks   179
   How rethinking time can help us choose a better future

7. The Lion-Man's Leap   215
   How synthetic biology could save vulnerable species from extinction

Kafka's Leopards   247

List of Cited Sources   249

Image Credits   275

Acknowledgements   277

Cliff swallow with shorter, blunter wings

INTRODUCTION

Throughout the twentieth century, the cliff swallows of Capistrano were emblems of constancy. Each spring, the birds would travel thousands of miles, from their wintering grounds in Argentina to their nests in the eaves of the Mission San Juan Capistrano in California, on exactly the same day each year: the 19th of March. Their return was greeted as confirmation that the cycles of the natural world would revolve forever, unyielding and true.

Cliff swallows are a small passerine bird, typically no larger than the span of a human hand. They wear a white triangle on their foreheads and bright rouge on their cheeks, over a dark cap and coat. Their wings are long and pointed, but instead of the barn swallow's elegant tuning fork, cliff swallows' tails are as squat and square as shovels. The birds dwell in cliff-face colonies of several thousand, where they build gourd-shaped nests out of mud.

Their reputation for fidelity to the calendar does not rule out an ability to adapt where necessary, however. Cliff swallows were originally distributed along the California coast, but the westward expansion of the United States in the nineteenth century provided opportunities to spread east. Woodland cleared for pasture created fields and ponds with plentiful insects to feed on, and with each new farm and town there were a host of cliff-like nesting sites in the shape of barns and buildings.

In particular, the birds learned to love roads. They followed wherever the highways led, building their pot-bellied nests under bridges and in culverts, in colonies of raucous thousands. Ornithologists began studying the social behaviour of roadside-dwelling swallows in the 1980s, and as the years passed they began to notice something strange.

Roads are typically dangerous places for birds, especially since they have come to be dominated by hulking SUVs. But as the cliff swallow population grew, unexpectedly the numbers killed by cars steadily declined. When the scientists compared the wingspans of roadkilled birds with those of the general population, they discovered a curious difference. Roadkilled swallows' wings were characteristically long and pointed, but in swallows that managed to avoid traffic they had become short and rounded – trowels rather than stilettos. This blunter shape made the birds nimbler, better at sudden vertical take-offs and the knife-edge turns that would take them out of the path of oncoming vehicles. The surviving birds passed their genes to their offspring. Short-winged swallows, it seems, were just better at playing chicken.

Evolution can have many different causes. A food source may decline, or a new one appear. Helpful mutations can bubble up in the gene pool. Populations of different species come together, spurring one another to adapt to competition or threat, or drift apart, to develop in isolation. All living forms are in some way a product of the combination of pressure and time. The particular pressure bearing down on cliff swallows' wings came from three-ton SUVs travelling at high speeds. In just a few decades, these emblems of unchanging nature had evolved, and the reason was us.

*

INTRODUCTION

We live in a world of wonders beset by the most profound calamity. While cliff swallows have found a way to live with at least one kind of human threat, worldwide half of all bird species are in decline. In 2019, a United Nations report predicted that, all told, up to a million plant and animal species could become extinct by 2100. But in the midst of this crisis, something else – subtly, quietly – is happening. Some organisms are changing, and with astonishing speed. Each of the main drivers of extinction – agriculture, urbanisation, the spread of introduced species, pollution, even climate breakdown – is also driving evolution.

For nearly 4 billion years, life on Earth has experimented with ways of being, sensing, moving and reproducing, finding ever-new shapes by which to meet the challenges of the moment. But the demands of life on a human planet are testing this ingenuity. On every continent except Antarctica, animals, plants and insects are altering their bodies and behaviours in response to the pressures of transformed ecosystems and a changing climate. Warming global temperatures expand the range of everything from corals and mosses to birds and butterflies. In cities, tall buildings, subterranean railways, and parks offer synthetic versions of cliffs, caves and waterways. Transoceanic shipping has knit back together continents that split apart hundreds of millions of years ago. Our fingerprints are everywhere: in birds that forget their songs, in city-dwelling spiders weaving new web patterns, and in elephants born without tusks to escape the murderous attention of hunters. Human civilisation is now the world's greatest evolutionary force.

As the conditions for life on Earth shift, with potentially devastating consequences, what can this widespread change teach us about how to shape a liveable future? In fact, there is much we can learn from evolution's flair for transformation, even where it bends under human influence — about issues as urgent and diverse as how to build sustainable cities, how to deal with pollution and how to address climate change.

Before we approach these lessons, though, a word of warning: the fact that humanity is now driving evolution — a force that has shaped life on our planet for 3.7 billion years — should not be seen as cause for celebration. Nor can we find in the fact that humans are driving evolution reason to feel complacent. Adaptation is not a panacea. The fact that many species are adjusting to the challenges of life on a planet dominated by us does not absolve us of responsibility for the biodiversity crisis; we can't just sit back and let natural selection pick up the pieces of our fractured ecosystems. For as long as we continue to alter the chemistry of the atmosphere and the oceans, to carve up the environment for roads and resources, and to flood air, soil and water with industrial toxins, then death on a massive scale will follow. For most species, the pace of climate breakdown and habitat destruction is too great; the more delicately fitted a plant or animal to its niche, the less likely it will be to adjust when that niche becomes untenable. Ecosystems of the near future will consist of the hardiest and most adaptable species. But even for many of those that are able to adopt new characteristics, there is a high probability that they will miss the target. Maladaptation, where a change ultimately leaves a creature more vulnerable, is a continual risk. Biotic change rarely happens in isolation. Ecosystems are intricately and precisely calibrated to the vast number of interdependent relations that coexist within them; even subtle changes to these relationships can, cumulatively, have severe consequences.

Nonetheless, change is happening, everywhere and all at once. Some adaptations will allow other creatures to cope with the demands of life on a human planet. In time, if they flourish,

INTRODUCTION

the changes instigated by us will reset the dial on the evolution of those species. In other words, what happens now could have profound implications, even for the distant future. Evolution is irrepressible. It is nature's restless genius in action, and when the very basis of life is so threatened, we ought to consider what it can teach us about making a better world. Evolution was once thought to be ponderously slow, a languorous unfurling over millions of years. But our impact has been extraordinarily rapid. In just 100 years, North American songbirds developed new wing shapes as the forests that once blanketed half the continent fragmented. In 1948, Paul Müller won a Nobel Prize for the discovery of the insecticide DDT, but even before the ceremony had taken place, houseflies were showing resistance to its effects. By the 1960s, mosquitoes were also resistant; by 1990, more than 500 species were immune to DDT. Since the 1980s, bighorn sheep in Alberta have responded to trophy hunting by growing smaller horns and bodies. Oceanic plankton in Venezuela's Gulf of Cariaco adjusted to a thermal niche half a degree warmer than the surrounding ocean within fifteen years. Every day, roughly 100 billion trillion antibiotic integrons – the genetic mechanisms behind bacterial evolution – enter the environment via human and animal waste, each with the potential to set off a fresh wave of microbial resistance.

In 2016, scientists at Harvard University designed an experiment to show just how rapid evolution can be under human pressure (albeit their subject was microbes, whose talent for reinvention far outstrips that of more complex organisms). The mega-plate experiment involves a table-sized petri dish divided into nine bands, containing increasing levels of antibiotic the closer they are to the centre of the dish. In the outermost band, there is no antibiotic; in the next there is about as much antibiotic as *E. coli* can tolerate; then ten times as much, then a hundred, and so on, until the central band, which contains 1,000 times as much antibiotic as the bacteria can survive.

A time-lapse video filmed by the researchers shows what happens when *E. coli* are introduced to the outer bands. The agar in which the bacteria swim is mixed with ink, whereas the bacteria appear as white. Initially, the microbes expand to fill the bands without any antibiotic, until they are arrested at the boundary of a band with a greater concentration of antibiotic than they can cope with. Then a single mutant, like a flare against the night sky, crosses the dividing line. Further mutations follow, competing with one another for space as they spread from band to band, until they reach the environment with 1,000 times as much antibiotic as *E. coli* in the wild can survive. The white bacterial colonies blooming in the inky agar look like sea-ice crystallising on the surface of a dark sea. A counter indicates that the journey to the lethal centre takes just eleven days.

The lesson of the animals, plants, insects and microorganisms responding to the pressure of living on a human planet is that change can happen remarkably quickly. This potential in other beings ought to be an example for us to follow. We are forcing nature to reimagine itself, and to avert calamity we need to do the same.

Fortunately, the capacity for change lies in every living thing.

Most of us have learned to think of biological change as residing in genes. But for the first half of the twentieth century, genes were little more than figments of the scientific imagination. The rediscovery of Gregor Mendel's pioneering work on heredity in plants, ignored on publication in the 1860s only to emerge again in 1900, was the foundation for the new science of genetics, a term first used in 1906. 'Gene' was coined three years later, but it was an idea without a physical form. No one knew what

INTRODUCTION

genes looked like, or even if they actually existed. For many scientists, genes were just a necessary fiction to describe the fundamental insight of genetic science: the steadfast transmission of traits from generation to generation. Yet – or perhaps as a consequence of their insubstantial form – they attained an almost mythic power. In 1944, Erwin Schrödinger declared the gene a kind of biological tyrant, both 'law-maker and executive power – architect's plan and builder's craft – in one'. The 1953 discovery of the structure of DNA by James Watson and Francis Crick, drawing on the crystallography of Rosalind Franklin, revealed the chemical substance that makes genes, but the rooting of the gene in the physical world only served to elevate it even further. Genes became the primary agents of evolutionary change, 'part physicist's atom and part Platonic soul', as physicist and philosopher Evelyn Fox Keller puts it.

Siddhartha Mukherjee writes that twentieth-century science was defined by three essential units: the gene, the atom and the byte. Of the three, the gene may have had the strongest hold on our collective imagination, in part because it was understood to possess elements of the other two: genes were repositories of the most fundamental information about organisms, on the one hand, and also the source of a mysterious energy, a drive to make things happen, on the other. Latterly, the popular understanding of the gene drew more on the language of computer science than physics. Genes were imagined as a kind of biological Central Processing Unit (CPU), both data store and command centre for making living things. Change, either via mutations, natural selection or genetic drift, was like an error creeping into the program. But advances in genetics have complicated this picture.

Genes combine to form a genome, the sum total of information encoded in a creature's DNA; the genome produces a phenotype, a cluster of traits – from physiology to behaviour

– that define the species (in the manner of the *Just So Stories*, from the leopard's spots to the rhinoceros's wrinkled hide). But a genome can seed a variety of phenotypes. Genes can be switched on or off, assembling in multiple different configurations of the expressed and the unexpressed, and each different genetic assemblage gives rise to the traits that represent a distinct phenotype. The sum total of an organism's phenotypes makes its phenome – a repertoire of forms, each one better- or worse-fitted to a particular set of circumstances. The name for the ability to call on this repertoire is 'plasticity'.

This inbuilt capacity to adapt is found throughout nature, from dandelions to daphnia (or water fleas, which can adapt their body size, the structure of their filter-feeding mouth parts, and even the rhythm of their life cycles to meet the demands of different environments). But plasticity is not simply a matter of consulting a library of latent body plans and behaviours; it is perhaps better understood as the outcome of a collaboration. What an organism is or becomes is not only defined by the information in its genes, but also through interactions with the environment. A wide range of epigenetic factors may act on the genome – flicking the switches that activate or silence a particular gene – to produce different phenotypes. Phenotypic plasticity is the capacity to meet the challenge of environmental change with an altered body or refashioned behaviour – it's what allows the Arctic fox, for instance, to exchange its white winter coat for pale grey or brown in summer; it's also what blunts the curve of a roadside-dwelling cliff swallow's wings.

Most of the adaptations we can see today are not (yet) evidence of genetic change, but expressions of plasticity, exploiting an inherent flexibility in what it means to be a cliff swallow, or a green and golden bell frog, or a Greenland caribou. The ability to respond to a changing environment means that,

INTRODUCTION

despite lacking the vivid red plumage of their cousins who light up the forests of Ohio, city-dwelling northern cardinals are no less successful in finding a mate. The bright forest birds get their colour from their carotenoid-rich honeysuckle diet, and rely on their flashing feathers to attract the opposite sex; the scarcity of honeysuckle in cities dulls the costume of urban birds, who have responded by loosening the inherited link between ornamentation and reproduction. It's why organisms as diverse as corals, butterflies and trees can expand their range, why woodland salamanders and dragonflies can change their body shapes, and why green sea turtles produce more female than male young as the climate they inhabit gets warmer. Plasticity means that when the larvae of the African clawed frog ingest microplastics, they can increase the length and mass of their gut to compensate for the loss of nutrients.

Alongside all this plastic adaptation, some speciation – where change extends to the genome – is already taking place. Mosquitoes in the London Underground, as well as in the New York and Chicago subway systems, have evolved a subterranean way of life, and in doing so have lost the ability to breed with their overground cousins. As Europe gets warmer, a subpopulation of blackcap warblers has evolved a new migratory route, heading for Britain rather than Spain. Even in captivity, their offspring exhibit a distinct flight orientation, hopping in the direction of the UK rather than the Iberian Peninsula, indicating that the change is genetic. Each of these changes represents the beginning of a distinct new species; millions of years from now, the descendants of the London Underground mosquito may be one of countless species that owe their form to the long-distant influence of *Homo sapiens*. For now, however, most of the change we can see is down to plasticity.

Plasticity may even be the reason a fish was able to walk on land for the first time. The Senegal bichir is a smallish, lozenge-shaped fish with a row of thin spines along its back and two tough, stubby pectoral fins. A native of the Nile basin, it possesses both true lungs and gills, and is the closest extant analogue of the first fish to make the transition to life on land. A bichir's pectoral fins are strong enough to propel it along a hard surface, and in 2014 a study examined how it would respond to terrestrial living. The results were astonishing. The shoulder bones lengthened and formed stronger connections to the walking fins, while the attachments between the skull and body weakened, making the animals' heads more mobile. The landed bichirs ended up looking more like the pioneering tiktaalik, which left the ocean 360 million years ago. All of this happened within the lifespan of a single fish, or around eight months.

In a world where environmental change that would once have taken thousands of years now unfolds in decades, the ability to summon another body or a different feeding behaviour or migratory habit when the conditions demand it has become vital. This applies to us, as well. *Nature's Genius* is about the pursuit of human plasticity. The same potential for change seen in water fleas or Senegal bichirs is in us too; in fact, we have been exploiting our own plasticity for as long as humans have existed, and continue to do so despite being insulated from nature by technology, medicine and agriculture. We may not need to alter our genes, but we desperately need to change our way of life, and the example of the living world could help us realise this. Designing things with nature in mind could revolutionise everyday life. Mimicking coral could create buildings that repair themselves and actually draw carbon from the atmosphere. Imitating the way urban species improvise a wild life in our cities could revolutionise how we deal with our waste. Bioplastics made from apple peel and shrimp shells could

INTRODUCTION

replace many synthetic plastics. Drawing on the chemical genius of microbes could power our homes and transform the food we eat. More fundamentally, learning to work *with* nature in making a better world could revolutionise what it means to be human. We pollute because we see ourselves as separate from the rest of the living world, but the way other creatures are learning to live with chemical and plastic pollution can suggest ways to reconnect with the world around us. Climate change is altering the many 'wild clocks' that regulate migration, breeding and blossoming, but learning to coordinate our time with nature's rhythms – to make time with a whole forest – could revolutionise our politics. Understanding better our impact on how living things think, dream and communicate can help us reimagine what it means to live and work together.

Most of the changes driven by human action are unintentional. But these unplanned interventions in evolution are bookended in time by very deliberate efforts to reshape the bodies of other beings. Our first efforts to assist evolution – the domestication of plants and animals – were the foundations on which modern civilisation was built. Our future may depend on gene-editing technologies to boost the adaptive capabilities of threatened species. Both practices involve rethinking what it means to be human. We, too, are part of the changing world. It's vital that we embrace our own capacity to evolve.

For much of our recent history, being human has meant thinking of ourselves as a species apart. The capacity for change has always been nature's genius – something that, for all our ingenuity, we seem to have lost. The world is ripe for reimagining. But so are we.

Life on Earth is changing; the question is, can we change with it? If so, then our curious Earth can be a place where all of life can thrive.

Mechta ('Dream'), the first silver fox to develop floppy ears

I

OPTIMUM DOG

How domestication proves change is possible

The coming of spring was once a wild and savage time. During the Roman festival of Lupercalia, priests dedicated to Faunus Lupercus, the lord of all animals, would grin beneath a smoking altar as blood was smeared on their foreheads. Naked except for the ragged skins of sacrificed goats, they would run through the city terrorising all those they met with gory goatskin whips, in a pantomime of the god they wished to appease. Farmers believed that Faunus would punish them if they claimed any part of the forest for their crops and animals without first placating him. Lupercalia was intended to drive the wolf from the flock. 'Wolves rove among the fearless sheep,' wrote the poet Horace. 'Good Faunus, through my sunny farm Pass gently, gently pass.'

On 4 January 1800, just a few days after the turn of the new century, a local nobleman discovered a relic of Lupercalia beneath the walls of Castello di Lucera, in the Puglia region of southern Italy (where the heel of the boot pierces the Adriatic). The Lucera bronzes – a collection of fourteen bronze human and animal figures, now in the Ashmolean Museum in Oxford

– once formed part of a ritual vessel meant to hold a wine offering to Faunus, and date from the early seventh century BCE. One human figure holds a raised shield; another blows into a musical instrument of some kind. The animals include a bull-like creature, as well as a goose, a dog, a sheep, and several goats' heads with elegantly curling horns. Perhaps the ritual to appease Faunus succeeded; nineteenth-century records show that the group may once have also included a wolf with a lamb in its jaws.

Without the wolf, what is left resembles a familiar rustic scene of people and the animals they have domesticated. But on closer inspection this familiarity seems misplaced. One nineteenth-century palaeontologist speculated that the bronze goats are in fact wild Alpine ibex; and the bull may be *Bos primigenius*, the great (and now extinct) auroch from which all species of cattle are descended. The animals flicker between wild and domesticated, forest and farm. And then there is the dog.

Dogs are our oldest companion animals. They have been fully domesticated for at least 14,000 years, but archaeological and genomic evidence suggests that the process of separating dog from wolf began much, much earlier. Fossil footprints left in the Chauvet Cave in France 26,000 years ago show a child running alongside a large canid that, from its size and gait, appears to have been more dog than wolf. A dog-like skull discovered in a cave in Belgium has been dated as over 30,000 years old, and genomic analysis of a mummified Siberian wolf indicates that dogs began to diverge from wolves as much as 40,000 years ago. By the time the Lucera figures were cast, dogs had been dogs for a long time. But the canine ornament none-theless seems to recall something of its wild past. 'I hear with ears that point upwards', writes the poet Anthony Vahni Capildeo in 'Dog or Wolf', their poem about the Lucera bronzes; whereas 'satisfaction curls over my tail'. Wolfish features mingle with those of the hound, the pricked ears of *Canis lupus* contrasting with the eager tail of *Canis familiaris*. 'Good lupo', as Capildeo says, shares a form with 'optimum dog'.

It is as if the whole millennia-long history of domestication, of shaping one species into another to serve our needs, is expressed in this strange solitary figure; a history of animal change that has led to the human planet we now inhabit.

One day perhaps forty millennia before this one, a grey wolf somewhere in Europe or Central Asia caught the scent of a carcass discarded by hunters. For those animals who could overcome their natural aversion to people, there were rich pickings to be had at the edges of the places where humans gathered. Over time this opportunism led to companionship, as wolves and humans learned to live alongside one another. Gradually, the wolves began to change. Each succeeding generation looked a little less wolfish than the last. Evolutionary scientists call this the commensal pathway, where domestication occurs without any forethought or planning. Cats (probably in North Africa), chickens (in Southeast Asia) and pigs (in Western Eurasia) followed this same route. But our ancestors also learned how to manage herds of animals for fur, milk or meat. Captive breeding led to more directed breeding. It's no coincidence that goats, sheep and cattle were all domesticated around the same time in the same part of the world: the Fertile Crescent in the present-day Middle East, where agriculture flourished around 10,000 years ago.

The process of domestication was the midwife for our human planet. The world we know was born when our ancestors began to deliberately mould other beings. There's tentative evidence of small-scale weed cultivation on the shore of the Sea of Galilee 23,000 years ago, and by 9400 BCE Neolithic people in the Levant were farming the so-called eight founder crops: emmer wheat, einkorn wheat, barley, peas, lentils, bitter vetch, chickpeas and flax. Domestication also changed what it meant to be human. Cities, and the flowering of cultures that followed, are inconceivable without crop species. Harnessing the bodies of other beings enhanced what human bodies were capable of: just as the stone tool lent hardness to human flesh and the spear extended the killing reach of the human arm, domestication allowed us to draw on the strength of the ox, the speed of the horse and the dog's acute sense of smell. But it also literally changed the human body. Working with animals and crops gave us stronger arms and denser bones than hunter-gatherers; it also made us shorter. New diets made possible by cultivation nourished our farming ancestors, while new diseases plagued them.

To appreciate what nature's wild genius can teach us about life on a human planet, we must first understand our earliest interventions in the bodies of other creatures.

For long after people had mastered the domestication of other species, the mechanism that lay behind it remained a mystery. We knew how to do it, but we didn't know why exactly it worked. The matter of how heritable characteristics passed from one generation to the next was a gap in Charles Darwin's theory of natural selection. It wasn't until the second half of the twentieth century that a daring, multidecadal experiment – begun in secret in Soviet Russia – provided the answer.

In 1866, Darwin had published *The Variation of Animals and Plants Under Domestication*. It defined for the first time the features common to all domesticated species: changes in size and physiognomy; the retention of juvenile traits such as floppy ears or a curly tail; alterations to pigmentation (particularly the appearance of a white star on the forehead); increased sociability, including a greater tolerance for human contact; and earlier sexual maturity, coupled with an extended breeding season. Collectively, this cluster of commonly held traits came to be known as domestication syndrome, or the domestication phenotype. Variation was also meant to provide the missing link in Darwin's grand theory, however, and here the great man faltered. He proposed that every part of the living body produced minute 'gemmules' containing the precise instructions for reproducing itself. When passed to the offspring, each gemmule contributed part of the blueprint for the whole body: heart gemmules would show how to build a heart, and limb gemmules how to organise the limbs. He called his solution 'pangenesis'.

Darwin's notion of pangenesis is an inglorious footnote in one of the greatest scientific careers in history. To modern readers, the idea that each part of the body produces a tiny version of itself will seem quaint, even touching. Darwin's contemporaries were equally unpersuaded. His book on domestication sold only 5,000 copies in his lifetime (*On the Origin of Species* sold as many in a matter of months).

However, unknown to either Darwin or his readers, the true solution to the problem of heredity had already been glimpsed, in the study of an Augustinian friar called Gregor Mendel in what is now the Czech Republic. Based on his study of pea plants, Mendel had recognised that traits such as height or colour were inherited from both parents, in pairs he called 'alleles'; but the expression of that trait was determined by the dominant allele in the pair. Mendel published his theory of inheritance in the same year as Darwin's theory of pangenesis, 1866, to an even greater show of indifference, yet it was rediscovered to acclaim at the start of the twentieth century. The combination of Darwin's natural selection and Mendel's genetics became what is known as the modern synthesis – the foundation of genetic science.

In the 1930s, Darwin's account of domestication syndrome fascinated a young Russian geneticist called Dimitri Belyaev, who would go on to devise an extraordinary experiment that revolutionised our understanding of how wild animals are transformed by domestication. But Stalinist Russia was a dangerous place to be a follower of Mendelian genetics. At least 5 million people died of starvation during the Great Famine that lasted from 1930 to 1933; several million more would perish in the famine of 1946–7. Stalin was desperate for scientific innovations that would stave off further food shortages and justify the agricultural reforms that had driven Russians to the brink.

In this febrile atmosphere, a charlatan saw an opportunity. Trofim Lysenko, the son of Ukrainian peasant farmers with only a correspondence degree in gardening, claimed to have devised a method of vastly increasing grain yields during cold weather by freezing seeds before planting. The science was wholly bogus and Lysenko fabricated his results, but a talent for self-promotion and political manoeuvring meant he caught the attention of Stalin, who lauded the 'barefoot professor' as a Soviet hero. Lysenko, for his part, jealously guarded his position by persecuting Mendelians as proponents of decadent Western ideas. The danger could not have been more acute: following Lysenko's furious denunciation of geneticists as 'saboteurs' at an agricultural conference in 1935, Stalin himself gave a standing ovation.

Belyaev was passionate about Mendelian genetics and fascinated by the effect that human environments had on other species. But Lysenko's grip on Russian science meant that he had to proceed cautiously. Outing yourself as a geneticist could be fatal. During a visit to Moscow in 1937, Belyaev's idolised older brother Nikolai, who had inspired young Dimitri to become a geneticist, was arrested and shot without trial. In 1940, the leading genetic scientist of the day, Nikolai Vavilov, was met on the street by four men in dark suits and taken to Moscow's infamous Lubyanka prison. Vavilov, who had collected more plant and seed samples than any other scientist of his generation in order that the terrible famines of the 1930s should never happen again, starved to death in 1943.

In 1939, Belyaev began working as a fur breeder at Moscow's Central Research Laboratory, and gained a reputation for breeding furs in extraordinary colours, from cobalt blue to lustrous pearl. In 1952, a secret visit to a fox farm in Tallinn, Estonia, which had succeeded in breeding tamer foxes over just three generations, planted the seed of an idea. By 1959, Belyaev was director of the Institute of Cytology and Genetics in Akademgorodok, a purpose-built research town near the city of Novosibirsk, 2,000 miles from Lysenko's thugs in Moscow. The stage was set for what would become one of the longest-running experiments in animal evolution.

In 1910, silver fox furs farmed on Prince Edward Island in Canada had sold for $2,500 a pelt. Their descendants had been introduced to Siberia in the 1920s as a source of revenue, and Belyaev's extraordinary success as a breeder provided the perfect cover for his experiment. If any of Lysenko's heavies enquired, he could claim he was looking for ways to add to the Communist Party's coffers by extending the foxes' breeding season. In fact, he wanted to test a theory that he could release a host of changes in their bodies by radically increasing the animals' exposure to human contact.

His ambition, Belyaev confided to a colleague, was 'to make a dog out of a fox'.

Akademgorodok did not have capacity for the experiment

he had in mind, so Belyaev commissioned his assistant, Lyudmila Trut, to establish a breeding centre at a fox farm in Lesnoi, over 200 miles from Novosibirsk. There she oversaw thousands of captive foxes, each pacing its own cage inside row after row of open-air sheds. The noise and the stench, she later recalled, were overwhelming. The method Trut used to identify the tamest individuals among the yowling horde was disarmingly simple. She would count her steps as she approached each fox in turn. The most tolerant pups – those that allowed her to get the closest before snapping or snarling – were selected to form a breeding pool (the scientists also ran a control line of unselected foxes).

Although the Prince Edward Island foxes had been farmed in Russia for fifty years, there had been no effort to breed them selectively. They looked and behaved like wild foxes, snarling at any approaching humans and retaining the same seasonal breeding pattern. Belyaev anticipated it would take fifty years of selective breeding for the experiment to yield any results; in fact it was more like fifty months. After just three generations the foxes had lost their fear response. One pup born in the fourth generation (called Ember by Trut) was the first to wag its tail; pups in the sixth generation became excited when Trut or others approached, licking the humans' hands and whining pitifully when left alone. Some of the eighth generation had curly tails.

Lysenko's grip on Soviet science was loosened when he was deposed as Director of Soviet Biology in 1964. In contrast, Belyaev's experiment was flourishing. By 1967, the evidence that human contact was radically changing the animals was so great that Belyaev commissioned a purpose-built farm at Akademgorodok. His enthusiasm for the foxes would sometimes overtake him: when he described the experiment to fellow scientists he would imitate the pups, pooling his eyes and curling his wrists in a begging posture.

The breakthrough came in 1969. All wolf and fox pups have floppy ears, which straighten out after a few weeks; but in one fox – which Trut named Mechta ('Dream') – they drooped into adulthood. Another pup had a white blaze like a star on its forehead: the classic mark of the domesticated creature in Darwin's schema. When he first met the pup, Belyaev was said to have exclaimed, 'And what wonder is this?'

At the same time, the foxes's annual breeding pattern was broken. Forty per cent of the females were breeding three times a year. Tests revealed that the domesticated animals also had half the stress hormones of the control line. These floppy-eared, piebald creatures who could produce multiple litters a year and actively sought human company were no longer like foxes anywhere else on Earth. It had taken just ten generations to turn a fox into something that looked a lot like a dog.

The changes continued. Between the fifteenth and twentieth generations, the foxes developed shorter legs and tails; some had pronounced over- or underbites. Their skulls became narrower, and the males began to look more like the females. In 1974, Trut moved into an experimental house with a pup she named Pushinka (meaning 'little ball of fur'), to test how the fox would cope when living with a human. After an unsettled first night, Pushinka slept at the foot of Trut's bed and would bark at strangers. The American Kennel Club contacted Belyaev about importing his foxes as pets.

Fox pups, like wolf pups, are not born with an aversion to humans, but they become more aggressive as they get older. Belyaev realised that delaying (or even switching off) this tendency towards aggression – effectively arresting the animal in a state of immaturity – triggered a host of other, seemingly unrelated changes in their body and behaviour. In a 1978 lecture to the International Congress of Genetics in Moscow, he named this process 'destabilising selection'. In a relatively predictable environment, natural selection will weed out extremes, creating a bias towards equilibrium between an organism and its environment that is maintained from generation to generation. But when a sharp new pressure emerges, that stability is thrown off and one change can trigger a host of unanticipated variations.

Later research has identified the likely physiological cause of destabilising selection. Neural crest cells are a form of stem cell associated with the hypothalamic-pituitary-adrenal axis, which underlies the production of stress hormones. They're also associated with the development of cartilage in the ears and tail, the tissue of the jaw and teeth, and pigmentation. By selecting for tameness, domestication introduces defects in the way neural crest cells develop these other traits as well, producing floppy ears, curly tails and so on.

Belyaev's experiment with silver foxes didn't reproduce exactly the way wolves became dogs thousands of years ago. That process would have been much slower, being less intensive and lacking the clear objective (the notion of 'dog') that Belyaev had in mind. But it did reveal the extraordinary degree of plasticity that lies within living things – a plasticity our ancestors exploited to create new species and lineages that would serve human needs. In doing so, they established a degree of mastery and control over other living things that had never been seen before. Aurochs became cattle; wild Asian mouflon morphed into sheep. In each case, destabilising selection was the catalyst: we drew the animals closer to us, and in a literal sense transformed them by our presence.

Belyaev recognised that domestication without close supervision can also lead to many undesirable characteristics. Domestication is, he stated, 'one of the greatest biological experiments'. It must have involved a great deal of trial and error to make the domesticated animals we now know. But in mastering the process, people turned themselves from a destabilising into an artificial stabilising influence, controlling selection to ensure that each new generation of cattle or sheep looked the same and was as productive as the last. Eventually, this emphasis on standardised animal bodies optimised for productivity would have extraordinary – even monstrous – consequences.

In 2014, a team of Swedish geneticists replicated Belyaev's experiment in a population of red junglefowl. Applying the same method of selecting for tameness, they found that after just three generations the birds lost their fear of humans, grew heavier more quickly and laid larger eggs. Junglefowl look like miniature chickens; the males, with their green and gold feathers and scarlet combs, resemble the cartoon cockerels in a children's picture book. This is perhaps unsurprising, given that modern broiler chickens are descended from a population of red junglefowl that lived in Southeast Asia around 8,000 years ago. But modern-day farmed chickens look very little like their picture-book counterparts. Like Belyaev's foxes, industrial broiler chickens have been bred to mature much quicker than their wild cousins, typically reaching adulthood at five weeks, when they are sent to slaughter. By this point the birds are grotesquely over-developed, with such massive breast muscles they have trouble walking and even breathing.

Broiler chickens are now three times larger than the wild red junglefowl from which their ancestors diverged. Most of this growth has happened in the past seventy years. A broiler chicken born in 1957 would be between a fifth and a quarter of the body weight of the birds we farm today. But the attitudes that made this possible are older. In 1916, an American farming textbook recommended that the animal farmer 'think of himself as a "manufacturer", for he too converted raw materials into valuable finished goods'. This shift in thinking transformed a wild bird into a standardised body plan, designed to yield the maximum product in the shortest possible time.

The industrial farm has become a uniquely potent evolutionary environment, and in evolutionary terms a grossly successful one: the standing global population of chickens – the number alive at any given point in time – is 22.7 billion, an order of magnitude greater than the next-largest bird population, the red-billed quelea (1.5 billion). They are now the largest population of a single bird species ever to have existed (at their peak, the passenger pigeons that once were so numerous they darkened the skies over North America numbered no more than 5 billion), and are present on every continent but Antarctica.

In due course, destabilising selection also destabilised our world. Today, farming is one of the main drivers of ecosystem change, replacing forests rich in life with fields full of cattle or crops. Domesticated species make up 60 per cent of all the biomass of land-living mammals on the planet. Nearly 40 per cent of global land surface is set aside for agriculture, forcing all other creatures to learn to live on the margins of a domesticated world. But could another domestication revolution restore balance?

I had junglefowl and chickens on my mind one bright May morning as I walked through Edinburgh's New Town to meet the environmental campaigner George Monbiot. The previous evening I had watched him give a talk in which he called industrial farming 'the most destructive human activity ever to have blighted the Earth'. In his book *Regenesis: Feeding the World Without Devouring the Planet*, he lays out the charge sheet against 'Big Farmer'. The expansion of grazing land is the world's greatest cause of habitat loss, responsible for 80 per cent of the deforestation of the past century. One-third of greenhouse gas emissions come from farming. Half the crops grown today are used to feed livestock whose populations have boomed in the past fifty years; the chicken population is five times what it was in the middle of the last century. The genetic diversity of the world's crops has collapsed to just a quarter of what it was in 1900, dominated by just four plant species (wheat, rice, maize and soy). The sprawl of farms into virgin rainforest like the Amazon threatens 24,000 of the 28,000 species known to be at risk of extinction.

I asked George what part domestication has played in the transformation of our planet.

'By far and away, the most destructive aspect of farming is livestock farming,' he replied, 'simply because it requires so much of the Earth's surface.'

Agricultural sprawl has replaced teeming forest, wetland and savannah ecosystems with the same blasted landscape. Planting and harvesting the crops we prefer means visiting an artificial catastrophe on the soil they grow in, clearing it of other plants and evicting most insects in the process; creating pastureland means exchanging rich ecologies for mile upon mile of ruminant-cropped grassland. Every lost acre of forest or wetland comes with a carbon opportunity cost: 1,250 kilograms of carbon that would otherwise be locked away is released for every kilogram of pasture-fed beef. By that calculus, George reckoned, four kilograms of beef (or a month of Sunday roasts) has the same carbon cost as a return flight from London to New York.

I mentioned broiler chickens. 'We kill 66 billion of these birds each year,' George said. 'The vast majority are produced in massive steel factories. There could be 50,000, 60,000, 70,000, even 100,000 chickens at any one time, all packed together in horrendous living conditions. They're just an industrial product, killed and plucked and chopped up, and fed on substances from all over the world – primarily soya from Brazil.'

The extraordinary quantities of nitrogen and phosphorus fertilisers involved have further consequences, he said. 'Since 2011, for six months every year, a belt of sargassum weed stretches from the Gulf of Mexico, down the coast of Brazil and right across the Atlantic to the coast of West Africa: a single, thick band of this weed.'

That dwarfs the Great Barrier Reef, I said.

'Oh yes, it's much, much bigger,' George replied. 'It girdles a quarter of the Earth.'

The band is fed by fertiliser run-off coming down the Tapajós, Xingu and Tocantins rivers, flowing into the Amazon and out to sea, and supercharging the massive sargassum blooms. 'Then the weed dies and rots,' George said, 'and as it does so, draws all the oxygen out of the water and creates dead zones' – vast areas of ocean where the lowest layers of the water column are so depleted that almost nothing can survive. It is now one of the greatest spectacles of the natural world, he explained; a collateral effect of mass-producing chicken and pig feed.

If the ills of industrialised farming are unequivocal, however, so are the needs it exists to serve. How do we reimagine animal and human life together when 8 billion people need food to eat and clothes to cover themselves?

It is often said the greatest impact an individual can have on climate breakdown is through their diet. A 2022 study forecast that rapidly phasing out animal agriculture could produce 'a 25 gigaton per year reduction in anthropogenic CO_2 emissions, providing half of the net emission reductions necessary to limit warming to 2°C'. In an entirely vegan world, food emissions would drop by 70 per cent, releasing huge areas of land for rewilding, which would massively expand the amount of wild biomass and create a vast new carbon sink. Levels of other greenhouse gases like methane and nitrous oxide, much of which is produced by animal farming, would fall precipitously.

Change of this order would require an extraordinary revolution in our relationship with the land and with other species. Farmers would need to massively scale down breeding programmes; the factory template of industrial agriculture would be replaced by 'rewilded agriculture', a patchwork of different methods designed to flourish in different contexts. It goes without saying that Western habits of consumption would be a thing of the past, a historical anomaly.

But for George, a life truly free from 'Big Farmer' lies in an even more radical departure from traditional ways of feeding ourselves. Soil is one of the richest environments on the planet. One square metre of soil can be home to hundreds of thousands of microscopic animals and thousands of species, many not yet known to science; a single gram can contain a kilometre of mycorrhizal filaments – the ghostly threads that tie together plants and fungi in a symbiotic covenant – and a billion bacteria. Intensive farming pushes fertile soil to exhaustion; climate breakdown has dramatically increased erosion. We pour great quantities of antibiotics, long-lived toxins such as pesticides, and even microplastics (because they make earth more friable and so easier to plough) into our soil. This profligacy seems astonishing, as George pointed out, given it takes up to a thousand years to make just one centimetre of topsoil.

Instead of intensive agriculture, he proposes that we look to the incredible inventiveness of the soil itself – or at least, the bacteria within it – to feed a world of 8 or more billion people without devastating what remains of the Earth's ecosystems.

Cupriavidus necator is a soil bacterium that has evolved to draw energy not from photosynthesis, but from hydrogen. It can be fermented using only carbon, water and nitrogen sourced from the atmosphere, and produces an edible golden flour that is 60 per cent protein. A pancake made from this microbial flour tastes deliciously buttery, George told me, and was indistinguishable from any other pancake he'd eaten. If we were to adopt microbial fermentation as our primary source of protein, it could replace, for example, soybean production while using only 1/1700th of the land. Every human that has ever lived has depended on photosynthesis to provide the food we eat. The emergence of agriculture was an attempt to harness this process for our ends. But if we could accept bacterial protein in our diet, then 'for the first time in human history', said George, 'we will have a staple food that did not arise from photosynthesis.'

A 2020 report by London-based think tank RethinkX has proposed that farming the microbiome could represent a second domestication revolution. An often overlooked component of the first domestication, it states, is the role of microorganisms. Bringing cattle under human control also brought the bacteria in their gut – which help the cow digest nutrients, grow muscle and produce milk – into our orbit. The first farms were probiotic by way of managing the large fauna in which the microbes lived. According to RethinkX, the second domestication will do away with mammalian intermediaries, replacing the cow's rumen – the muscular system of stomachs that allows them to digest grass – with a 10,000-gallon, temperature-optimised steel fermenting chamber to produce the proteins, carbohydrates and fats we currently get from dairy products.

In some respects, the notion of first and second domestication revolutions is a simplification of a much more complex and nuanced process. Archaeologists have suggested that, initially, Neolithic animal husbandry was focused entirely on meat, and many generations passed before a 'secondary products revolution' saw farmers also exploit their animals for milk, wool or hide. The changes initiated by industrialised farming are radical enough to represent yet another distinct revolution.

Where some see a revolution, however, others just see history repeating itself. 'The first domestication allowed us to master macro-organisms,' the RethinkX report claims. 'The second will allow us to master micro-organisms.' Food researcher Julie Guthman suggests that this emphasis on hyper-efficient production simply replicates the detachment of bio-industrialised agriculture, with its light- and temperature-controlled hatcheries and antibiotic-boosted growth schedules. Industrialised agriculture is driven by three principal aims, she states: to improve efficiency through continuous production; to standardise animal bodies so they fit the models of greatest productivity, shaped to fit seamlessly to the milking machine or slaughterhouse disassembly line; and to reduce the risks this involves, for instance by flooding those standardised bodies with antibiotics. Shifting focus to the microbiome simply exports these priorities while adding a fourth: to make other beings play the role of livestock and land at once, making microbes into 'living factories'. Microbes cultivated to manufacture protein would be both field and herd.

Images of battery-farmed microbes might not engender the same empathy as chickens or pigs raised in such depraved conditions. Microbes can't be said to suffer the way mammals or birds do, and we might therefore suppose that shifting agriculture to the microbiome would resolve the contradiction inherent in claiming to love the world while sequestering so much of it to feed our appetites. But if, in doing so, we simply perpetuate the same attitudes that produced the industrial farm towards the rest of the living world, what really will have changed? On this basis, a farm that harvests protein from microorganisms – standardised and optimised in laboratories, and contained in bioreactors – might look very like one that does so from cattle and chickens.

We need to feed ourselves in ways that don't contribute to the collapse of biodiversity and the breakdown of a liveable climate; that much is clear. But a revolution in how we do so needs to include a transformation in ourselves. George sees the potential for this in rewilding the land freed up by microbial farming. 'If we allow very large parts of the Earth's surface to rewild,' he told me, 'not only will that lead to a massive new fluorescence of nature and drawdown of carbon from the atmosphere, but also to a reconnection between human beings and the natural world. We need it spiritually as well as in practical terms. We could see a great reconnection taking place.'

Our ancestors valued their livestock because, without them, they would starve. The hope and fear expressed in the Lucera bronzes were intimately bound together with the fate of the herd. Somewhere along the way, that link was cut and animal life became a commodity. But 'what would happen', writes food activist Michael Pollan, 'if we were to start thinking about food as less of a thing and more of a relationship?'

Of course, we all have a relationship with food even if we don't think of it in these terms. Many of us invest a great deal in keeping this thought at a distance. 'Until I was twenty-six years old, I had never looked at a piece of meat,' writes Amber Husain in *Meat Love*. 'I had seen and eaten a lot of them [but] all I remember witnessing is the juicy abstraction of "meat".' 'Meat' is a cleansing concept, at least for those of us not involved in producing it; it washes away the grossness of flesh and tidies away the uncomfortable thought of the life it once sustained. Would it eventually be the same to look at a pile of protein-rich microbial flour and witness the tasty abstraction without the unsettling thought of the germ that made it? Everything we eat that comes from animals has undergone some kind of transformation – fermentation, pasteurisation, or a kind of conceptual magic that turns flesh into meat – but an aura of the natural survives these processes. It's a stretch to imagine that supermarket shelves might one day be lined with products featuring cheerful images of *Cupriavidus necator* in place of happy cows in green fields; but much less of a stretch, I think, to imagine that eating its protein would become something unremarkable.

But to allow the produce of the microbial domestication revolution to dissolve into abstraction would continue the legacy of industrialisation. To connect in a new way with what we farm and what we eat means also connecting in a new way with our sense of who we are. Ten thousand years or more of mastery over other species has fixed a certain idea in our minds: that domestication is something we do to other creatures. But that is far from the whole story.

A few weeks after meeting George, I travelled to London. It was early summer and the city was crowded with visitors gathered to celebrate the Queen's Platinum Jubilee. The atmosphere was festive, albeit in an ersatz way: plastic Union Jack bunting hung in shop windows and the porters at King's Cross station wore paper crowns.

Having reigned for seventy years and reached the age of ninety-six, Elizabeth II had become the third longest-reigning monarch in world history. The length of service was exceptional, but her age less so. Although above the average for her generation, such longevity was in keeping with the trend for longer life that has risen exponentially over the past century or so. Between 1891 and 2012, average life expectancy for women in the UK rose from 48 to 82 (for men, from 44 to 79). There are many reasons for this – chiefly improvements in healthcare and housing – but it comes down to the simple fact that we are as responsive to changes in our environment as any other species.

I hadn't travelled to join the jubilee celebrations, however. I was on my way to London's Natural History Museum to meet Predmosti and Spy, two individuals who I hoped could show me how changes to where and how we live also change what it means to be human. I wanted to understand how, as well as shaping many other species, domestication is responsible for making modern *Homo sapiens*.

They were facing each other across a crowded gallery when I arrived at the museum. Spy, short and robust, seemed the more relaxed, his rough hands clasped behind his back, feet squarely planted. A wry smile played across his face, his eyes crinkling beneath bushy brows. Predmosti, who was taller and more gracefully built, was far less welcoming. Tension ran through his whole body, which was half turned away from Spy, his torso twisting at the waist. His gaze was wary, as if it was a surprise to meet his cousin, here of all places. Nothing was said, but something deeply felt passed between them. Other than their tattoos and body paint, and a band holding back Spy's hair, both were entirely naked.

Predmosti and Spy are the work of Dutch 'paleo artists' Adrie and Alfons Kennis, who run the go-to studio for hominid reconstructions. Since 2006, working from Adrie's home in Arnhem, their sculptures of early hominids, Neanderthals and Neolithic *Homo sapiens* have been exhibited in museums around Europe. The distance of tens of thousands of years collapses in the presence of these impossibly lifelike, unexpectedly familiar reconstructions. In their recreation of the famous Tollund Man, whose body was preserved in a Jutland peatbog for seven centuries, his peat-stained sleeper's face becomes an open grin. Nana and Flint, sculptures of a Neanderthal adult and child, have an irrepressible vitality. Nana smiles indulgently at the small boy whose arms are flung around her waist, his hair standing up like a cartoon Einstein's, emphasising the mix of shyness and intelligence in his bright, dark eyes. What separates them from us recedes behind the sharp recognition of a grandmother's pride in a beloved grandchild. It's only belatedly that you begin to notice the physical differences: the heavy brows, flat cheekbones and receding chins; the short, stocky bodies.

Most remarkably, their reconstruction of Lucy – the partial *Australopithecus afarensis* skeleton discovered in Tanzania in 1974, which proved our ape-like ancestors walked before they had the brain capacity to develop tools – meets our gaze steadily across the gulf of more than 3 million years. Her body is covered in thick, dark hair and her features are simian; but there is the ghost of a connection in her eyes as they meet ours. The Kennis brothers' success goes beyond the individuality they imbue in each sculpture; in Nana's contentment, Flint's curiosity and Lucy's stillness there is a kind of welcome. They seem to recognise *us*.

Both Spy and Predmosti are based on the remains of individuals and named for the places where they were discovered: a 40,000-year-old Neanderthal found at Spy in Belgium, and a composite of a 30,000-year-old *Homo sapiens* skull from Předmostí in the Czech Republic and a headless skeleton found in Wales. Standing between them, it was difficult to remember that I was looking at models of different species.

The reality of *Homo neanderthalensis* is a long way from the brutish stereotype that emerged following the first discovery of a partial Neanderthal skull in 1856. A recent study of Neanderthal physiognomy has confirmed that they were capable of speech. Other research has established that they were skilled stoneworkers, wore clothes (possibly even shoes) and furnished their homes with plant-matter beds and animal-hide floor coverings. They mixed pigments – favouring a darker palate of black, red, dark brown, and grey – with which they decorated seashell ornaments and the walls of their dwellings. They adorned themselves with bird feathers, and cared for the sick and disabled. The oldest art we know about was made by Neanderthals, as was the world's most ancient structure, a ring of broken stalagmites assembled in Bruniquel, a cave in southern France, 174,000 years ago.

I saw all of this shared humanity reflected in Spy's smiling eyes. But what really joined me with him was our shared capacity for change. From the first hominid to now, through each new expression of what it means to be human, the common thread has been an innate plasticity – an ability to adapt to new pressures and opportunities. As the only surviving species of *Homo* we can sometimes think of ourselves as the finished article, the completed human. But viewing the history of our species as a line that leads directly and finally to us is a teleological trap. It not only keeps us from connecting with the non-human world in the way George Monbiot spoke about; it stops us from realising what we can be.

In the 1990s, an expert in diabetic medicine called Edwin Gale wondered why diabetes, a disease that was rare in the nineteenth century, had suddenly become so prevalent in the twentieth. It wasn't the diseases that had changed, he concluded; it was us. Our genes respond to different environmental pressures: lifestyle changes, such as from a mobile to a sedentary way of life, and the emergence of new factors like medicines or the modern city can lead to different forms of gene expression (or phenotype).

In *The Species That Changed Itself*, Gale describes three main changes in the human phenotype – transitions that correlate to the three major shifts in lifestyle throughout human history. Prior to the emergence of agriculture, hunter-gatherer communities ate less fat and more protein than we do today, and hardly any salt. The Palaeolithic phenotype was characterised by low blood pressure, powerful legs and lighter bones. Predmosti is long-limbed and lean; his people would have had the physiques and the stamina of long-distance runners, able to chase prey for hours until it collapsed from exhaustion.

As our relationship with other species changed, so did our bodies. The next phenotypic transition arrived with the first farms. Settled communities gave up the chase, and the bodies it had fashioned; the agricultural phenotype was characterised by stronger arms rather than stronger legs, with muscles developed through heavy labour and a denser skeleton on which to carry them. Our faces changed as well: a 2017 study found that the skulls of preindustrial farming societies had weaker jaws compared to their Palaeolithic ancestors, a consequence of eating softer foods like cheese. Human faces became narrower and more oval.

The genetic mutation for lactose tolerance gradually spread through European cattle-herding communities from about 11,000 BCE, a dietary advantage that also linked to reproductive success: carriers would produce offspring that were nearly 20 per cent more fertile than those without. Sedentary lifestyles made us less mobile in some ways, but it also led to more travelling between settlements, creating novel opportunities for diseases to spread and evolve. Visible alterations to our bodies were accompanied by invisible changes to the microbial communities in our gut and to our immune systems. Living in close proximity to animals and their parasites also introduced humans to new forms of disease.

Domestication did not just have a radical effect on the bodies of certain plants and animals, it also altered what it was to be human. Exchanging a nomadic lifestyle for a sedentary one changed our ancestors' relationship with place. Land became territory to be owned and defended, and stories, songs and myths arose to reflect this. Working the land meant they lived differently in their bodies, with the pain of backbreaking labour replacing the exhaustion of the chase. The focus of their fears shifted, with the stampeding herd or the predator's sharp tooth and claw superseded by drought or flood.

The next phase – the consumer phenotype – emerged in two stages, according to Gale. The first was due, in the West, to a levelling of social inequalities between 1870 and 1950. Social reformer Edwin Chadwick's 'sanitation revolution' recognised the link between poor living standards and the spread of diseases like cholera, transforming the life prospects of London residents in the mid-nineteenth century. This phenotypic transition even involved a change in human body temperature. A 2020 study found that the average oral temperature of US citizens has fallen by 0.03 degrees per decade since the mid-nineteenth century, most likely due to a population-level drop in inflammation.

The first stage of the consumer phenotype was just a prelude to the second, however. After 1950 – broadly coinciding with the reign that was being celebrated across London at the time of my trip to see Predmosti and Spy – our consumption of natural resources accelerated stratospherically. In the postwar era, fossil fuel consumption increased eight-fold; the number of people living in cities rose from 750 million to 4.2 billion. Again, we changed. We are, on average, twenty centimetres taller than people born in the first half of the twentieth century. Modern urban living has also made us significantly more prone to conditions such as obesity, myopia and dental impaction because of our narrower faces. (Gale notes that domestic animals may point the way here: some have lost their third molars, and already one in three people today are born without wisdom teeth.) Antibiotics and modern sanitation have deprived us of microbial communities that evolved over millennia to help regulate our immune systems, giving rise to autoimmune diseases such as diabetes and multiple sclerosis. The modern age of plastic has also been a time of profound human plasticity.

More than anything, Gale argues, 'the consumer phenotype was founded upon food abundance'. Nineteenth-century farmers' yields were limited by the scarce quantities of nitrogen they could extract from the soil or guano. But the invention of the Haber-Bosch process in 1910, which fixed atmospheric nitrogen into ammonia, offered a new and plentiful source of artificial fertilisers. Synthetic nitrogen now feeds upwards of half the global population. Its consumption rocketed from 46 million tons per year in 1965 to 190 million tons in 2019 – an ammonia mountain raised every year to feed a hungry world – dramatically increasing crop yields and nourishing upward trends for height, body mass and life expectancy with each succeeding generation. As with domesticated animals, faster growth is coupled with earlier sexual maturity: children now enter puberty four years earlier than they did a hundred years previously.

All of this prosperity and living in close proximity has been imprinted on our bodies. As we made our world a human planet, it was also shaping us.

Not everyone has been shaped in the same way, however. Privilege matters when it comes to gene expression. In some areas of Sub-Saharan Africa, people are shorter than they were fifty years ago. Gale notes that, fifty years after the 38th parallel divided North and South Korea, children in the south were thirteen centimetres taller and seven kilograms heavier than those in the north. 'A line on the map', he observes, 'had translated into a difference in biology.' As climate breakdown worsens inequality for many millions, we should expect its effects to also register on our phenotype.

The Palaeolithic phenotype existed for 95 per cent of human

history; the changes that followed were dramatically accelerated by domestication. But prior to this, domestication had an even earlier and more essential role in determining what it means to be human. In 1978, during his address to the International Congress of Genetics in Moscow, Dimitri Belyaev made in passing an astonishing remark. The destabilising selection he had observed turning wild foxes into tame quasi-dogs, he said, 'can also apply to humans'. Before we domesticated the world, we first had to domesticate ourselves.

Modern humans emerged between 350,000 and 260,000 years ago, but palaeontological evidence suggests that the first signs of domestication syndrome – smaller faces and a reduced brow ridge – were present in our ancestors as early as 315,000 years ago. For the past 2 million years we have been getting lighter as our bones have lost density, with an especially marked drop in weight as *Homo sapiens* emerged. The size of our teeth has diminished by around 1 per cent every 2,000 years for the past 100 millennia; 10,000 years ago, that rate doubled. The faces of men today are much more similar to those of women than at any time in the past 80,000 years. Domesticated animals also have smaller brains; despite growing steadily over 2 million years, ours began to shrink 30,000 years ago. 'The differences between modern humans and our earlier ancestors', writes primatologist Richard Wrangham, 'look like the differences between a dog and a wolf.'

As with Belyaev's foxes, the evidence points to a single cause: selection for tameness. Our ancestors learned that cooperation was an advantage in a savage environment. At some distant time, early humans learned to suppress the instinct – still vivid in other primates today – to lash out at strangers. We learned to coexist and to cooperate, and in doing so initiated the same suite of physiological changes found in every other domesticated species.

Modern humans retain a juvenile curiosity – an openness to new experiences – into adulthood. Many of us now live in communities numbering millions.

As I looked between Spy and Predmosti, something seemed to shift in their expressions. The wariness I had thought was there resolved into what looked like an understanding – a trace, perhaps, of that first hominin venture in collaboration, hundreds of thousands of years ago. 'All the world began with a yes,' writes Clarice Lispector. We changed ourselves by working together, and we continue to be moulded by the environments we create. Recalling the lessons of this plasticity will be key to our reconnection to nature, and to learning how we can live well in the world to come.

Dimitri Belyaev's experiment in how to make a dog proved that doing nothing more than exposing other creatures to human presence can have a profound effect, even leading to the emergence of wholly new species. But it works both ways: the first step in reimagining animal and human life together is to recognise that other beings have the potential to transform us too.

I left the National History Museum and stepped out into the festive atmosphere of the sun-drenched city. Children played with parents on the green outside; one girl turned cartwheels. Beyond the wall was the constant hum of the capital. London teemed with people whose bodies differed not only from those of Spy and Predmosti, but also from the bodies of Londoners who had walked the same streets only a few generations ago. Yet the thread running through us all began with the first experiments in self-domestication – in saying yes to others. Cooperation leads to change. And in the right conditions, radical change can be only a few generations away.

Magpie nest made out of 1,500 metal spikes

2

THE LIVING CITY

How urban evolution can teach us to build sustainable cities

The Krakatau eruption, on 27 August 1883, was the loudest sound ever heard by human ears. Three thousand miles from the explosion, people on the island of Rodrigues, in the Indian Ocean, reported hearing a low rumble 'like the distant roar of heavy guns'. Directly beneath the volcano the noise reached 310 decibels, loud enough to cause instant death. The shockwave travelled around the world three or four times, and then: silence.

Much of the former island of Krakatau had disappeared. Only a crescent-shaped fragment of its southern end remained, smothered in a forty-metre-thick layer of pumice. The remnant, called Rakata, was completely sterile. Nothing had survived the Hadean conditions of the eruption. In place of the riotous cacophony of a tropical forest, there was only the wind and the waves.

The first creature to arrive, when the pillar of superheated dust had settled and the raw ground cooled, was a tiny spider. Nine months after the eruption, a French expedition found this pioneer, spinning its web in an otherwise barren landscape. By 1889, insects had colonised the island, blown across the Sunda Strait or carried by the waves on fallen branches. Ten years later, monitor lizards swam over from neighbouring Java or Sumatra; by 1919, there were owls and flycatchers. Each new species brought its own retinue of parasites or a deposit of seeds in its guts. Fifty years after the blast rendered it a tabula rasa, Rakata was blanketed by thick forest. Life, it seems, cannot resist an island.

New islands are not as rare as you might think. In 1930, Anak Krakatau (or 'child of Krakatau') emerged from the volcano's sunken caldera. At least thirty islands have been born since the beginning of the twentieth century, including Surtsey, off Iceland, in 1963, and Hunga Tonga–Hunga Ha'apai, two islands in the Pacific connected by a volcanic bridge, in 2014. Many dissolve after a relatively short time, but some manage to cling on. Subsequent eruptions have continued to feed the growth of the child of Krakatau.

In fact, by far the greatest number of new islands are born on land. Cities, too, are islands – places of refuge where life congregates and thrives – and like islands they can emerge astonishingly suddenly. From Songdo, a 'city in a box' built from scratch on cleared ground in South Korea between 2000 and 2015, to the still-incomplete $100-billion Forest City in Johor, Malaysia, new metropolises have arisen across the globe like atolls of concrete and glass. Since 1949, China alone has built at least 600 new cities, each with a population numbering in the millions. The environmental effects are typically devastating, involving razing forests or reclaiming land from the ocean. But cities also, in a sense, replace what they destroy. Cities occupy just 0.5 per cent of global land, but they mimic nearly every environment on Earth: skyscrapers and tall buildings replace hills and cliffs; metro lines stand in for cave systems; reservoirs, ponds and culverts are substitutes for natural waterways. Cities offer an alternative to lost habitats, with plentiful food and shelter. The urban heat island effect – whereby the excess heat generated by people and machines can raise the temperature several degrees above the surrounding environment – even provides some cities with their own microclimate, attracting plants and animals like moths to a flame. Each city is both island and archipelago, a patchwork of niches that different organisms can exploit. And the networks of trade, culture and information that connect them – as well as the increasing homogeneity found in cities as distant as Minneapolis and Mumbai – mean they also form a planetary archipelago.

Islands shape life, as well as host it. In 1964, the largest earthquake ever recorded in North America raised up several islands in Prince William Sound, in the Gulf of Alaska, in a matter of minutes. The eruption created a series of ponds, scooping up ocean-going three-spined stickleback fish and confining them to an environment increasingly diluted by rainfall. Within fifty years, the sticklebacks had adapted to a freshwater lifestyle.

Like islands, all cities, both new and old, are evolution engines – hothouses of plasticity and speciation, where a host of creatures learn to adapt to city life. Urban snails in the Netherlands have lighter-coloured shells, to cope with the heat island effect; urban lizards in Los Angeles have larger scales for the same reason. Orb-weaving spiders from Australia to Belgium are making tighter, denser webs to compensate for the fact that there are, in general, fewer insects to catch in cities. Fish called creek chub in Raleigh, North Carolina, are changing their body shape to cope with faster currents in urban waterways, while house finches in Tucson, Arizona, have developed larger beaks with a stronger bite as a result of foraging sunflower seeds from urban birdfeeders, which are harder to break than food sources in the surrounding desert. Some species seem to swap traits: urban bridge-dwelling spiders have evolved an attraction to artificial light to catch insects, while ermine moths have lost their attraction to light entirely. City birds around the world are changing their tune: in Melbourne and Sydney, silvereyes sing in shorter bursts to cope with urban echoes, while the dawn chorus begins earlier along the stretch of the Jarama River that passes Madrid airport, to avoid air traffic noise.

In *The City We Became*, N. K. Jemisin's fantasy novel in which New York City itself has become sentient, there is a sound so faint only a few can hear it – 'beneath the others, the pillar supporting them, the metronome giving them rhythm and meaning: breathing. *Purring*.' If we watch and listen closely, our island cities are also alive: respiring not just with the tick of day and tock of night, the oscillation of work and rest, but with countless small lives, fitting themselves as we do to the demands of urban living.

All cities, with their scattered neighbourhoods and districts, and roads flowing between them like tarmac straits, call islands to mind. But nowhere is this more evident than in Dutch cities like Leiden, where the network of canals makes each bridge an isthmus and every block a distinct landmass.

Early one December morning, six months after my visit to London, I stood on the freezing deck of a North Sea ferry and watched the sunrise over the Dutch port of IJmuiden, a wash of cool pink and gold, as the biting air nipped my cheeks. I caught a bus to Amsterdam, then a southbound train to Leiden. It was bitterly cold. The weather report had said the temperature would be −3°C but would feel more like −8°C.

THE LIVING CITY

Frost furred the edges of the canals and the railings of the bridges, while the water below was glazed with frazil ice. Clusters of bicycles lined the narrow, cobbled streets, huddled together as if for warmth, their handlebars interlocking like the branches of winter hedges. I pushed my hands deeper into the pockets of my coat. The sky, though, was a deep, soul-lifting December blue, and the walls of the city were covered in poetry. Since the early 1990s, over a hundred murals called Muurgedichten, or 'wall poems', have been painted in Leiden. The poems are in multiple languages, by writers from Marina Tsvetaeva to Charles Baudelaire. I stopped on one corner, where a creamy white wall carried Shakespeare's melancholy Sonnet 30. 'I sigh the lack of many a thing I sought,' the poet complains. 'And moan th' expense of many a vanish'd sight.' I hoped it wasn't a foretelling of the day ahead.

I had taken an overnight ferry to the Netherlands in order to visit Menno Schilthuizen, an expert in urban evolution who had promised to take me on a tour of Leiden to see the process in action. He had suggested that we meet at his favourite café, located in a secluded courtyard behind the squat bulk of a Gothic church. Walking inside was like entering a domestic painting of the Dutch Golden Age, something by Johannes Vermeer or Pieter de Hooch: red stone floor, dark roof beams, and a fire burning in the brick-lined hearth. There were carved oak benches, and the whole space was illuminated by candlelight.

Menno stood up as I came in. He was tall, with short grey hair and a beard like the ice that furred the pavements outside, and he was wearing a snowy white rollneck jumper; but his greeting and handshake were warm. He invited me to join him in a mezzanine booth raised several feet from the ground – our own little island above the rest of the café.

Menno told me that he had begun studying observable evolution in snail populations in the mountains of Crete. 'You can see where evolution has taken place with almost every step,' he said. The snails moved so slowly that every hundred metres or so he found a distinct population with a different shell shape. This led him to study snails in urban environments (Menno was part of the research team that discovered the heat island effect was lightening their shells), and to the realisation that cities are hotspots for evolutionary change.

'We are nature's ultimate ecosystem engineers,' he explained. Where industrial agriculture has transformed much of the surrounding land into a sea of monocultures, we raise our cities like small islands of biodiversity.

'But there are also islands within cities,' he went on, and this environmental diversity is often the most important factor when it comes to how other species adapt. A patch of soil saturated with heavy metals represents an opportunity for plants that have adapted to thrive in polluted ground; an urban park is a green oasis in a desert of concrete. These sudden shifts from one environment to another create impassable barriers for many plants and animals, limiting gene flow. Rather than refreshing the genome, each local population consolidates its own distinctive features within its own 'island', a situation that could presage the emergence of a whole new species (as has already occurred with mosquitoes in the London Underground and the subways of Chicago and New York).

In some recorded cases, animals that occupied a place before the city was built have been marooned by urban development: white-footed mice, which lived contentedly on the island of Mannahatta long before the arrival of European settlers, found they were fragmented into distinct populations restricted to parks as New York City grew up around them. The same phenomenon applies to New York's dusky salamanders (as well as red-backed salamanders in Montreal and fire salamanders in Oviedo). Birds tend to be less constrained (in fact, great tits living in urban parks in Barcelona have shown even greater genetic diversity than their cousins in the surrounding forests), but less mobile species can become profoundly isolated. Some creatures have evolved to take advantage of the cave-like dwellings we humans tend to erect in cities; Menno said that certain urban spider populations can be unique to particular apartment buildings, and even to particular rooms. A study in Germany has found a similar parochial tendency in cockroaches.

Menno distinguishes between processes he calls 'soft' and 'hard' selection. 'In soft selection, the adaptation makes use of mutations that are already present in the population,' he explained. 'There's a lot of variation in populations that is at any moment neutral. But when conditions change, then suddenly it turns out there are already variations present which allow certain individuals to survive better than others.' The mice, birds, spiders, salamanders and cockroaches that have discovered a homely niche in parks or cellars did so by first discovering some latent potential within themselves to thrive in different corners of our cities. Hard selection, by contrast, involves a new mutation, such as that which occurs in some urban swans, where variation in the DRD4 gene – the so-called daredevil gene that is also present in downhill skiers and snowboarders – makes them bolder and more tolerant of their human neighbours.

We finished our coffee and wrapped up to face the punishing cold outside. Menno made a path through a tangle of narrow alleyways, passing tall and lean Dutch houses with steep pitched roofs and walls of dark brick. Along the canal, some of the older buildings listed at alarming angles, making rhomboids of windows and doors. Our walk had a meandering, improvised quality. Every so often, Menno would stoop to examine a tiny, urban-adapted plant sprouting in a pavement crack, or pause to observe a citified bird on a sloping roof. Everywhere we looked, the unofficial city was thriving.

Menno pointed to a blackbird on the rooftop across the street. 'As far as we know, this is one of the first birds to really form an urban population,' he said. Blackbirds first began occupying cities as far back as the sixteenth century, overwintering in the warmth of urban heat islands before giving up the habit of migration entirely and establishing themselves as permanent residents. Settling in the city didn't mean that the species stood still, however; as well as being non-migratory, urban blackbirds (or *Turdus urbanicus*) have developed shorter beaks and more high-pitched songs than their forest-dwelling counterparts. They breed earlier too, and have a decidedly more relaxed attitude to humans. All of these changes are encoded genetically.

Menno continued to lead me through the old city, towards Burcht van Leiden: an eleventh-century stone keep built at the point where the Oude Rijn and the Nieuwe Rijn, two tributaries of the Rhine, meet. Perched on raised ground, it looms over the surrounding streets like an artificial island.

We followed a path around the outside of the keep, which was strewn with fallen leaves glazed with frost. The mud was frozen into corrugated ridges. Menno suddenly crouched down by a pile of leaves and came up with a small, yellow snail. It was still alive, he said, despite the cold, and had been attempting to hibernate in the roots of a tree. 'Poor guy,' said Menno. 'They can usually dig up to thirty or forty centimetres into the soil. You wouldn't think they could do that.'

Shell colour in snails, he explained, is entirely genetic, like our hair or eye colour. Some feature of the environment was selecting for lighter-coloured shells, and the most likely cause was heat. In summer, the resulting albedo effect (whereby lighter-coloured surfaces reflect more of the sun's heat) would allow this snail and others like it to cope better with high temperatures than its darker, pink or brown cousins. It looked like a tiny sun in his black-gloved hand.

My walk with Menno taught me a series of lessons about how urban evolution can help us reimagine cities that are more sustainable, and more hospitable to all life. Cities often embody the worst excesses of our human planet, consuming vast resources and spewing out pollution. But by paying attention to the non-human city-dwellers who have learned to make the urban environment work for them, we could transform cities into beacons – bright spots showing the way to a better future.

Menno's snail – *Cepaea nemoralis* – illustrates the first lesson of urban nature: change your form. For urban blackbirds, this might mean changing beak shape; for urban spiders, it may mean weaving new web patterns. For urban humans, it could mean changing the form and the materials that make our buildings.

The way we build is immensely wasteful. Each year we manufacture 4 billion tons of cement, a key ingredient of concrete, extracting massive quantities of sand, minerals and aggregate, consuming billions of tons of fresh water, and producing 8 per cent of global carbon emissions. For some, the cliché of the urban jungle suggests an alternative: timber replacing concrete, even in the construction of skyscrapers. The world's tallest wooden tower is Norway's 85-metre Mjøstårnet building, but the city of Winterthur, in Switzerland, is building a timber skyscraper over 100 metres tall. Building with wood would not only shift construction to renewable sources, it could also make cities and settlements more responsive to climate breakdown and sea level rise: a wooden house can be deconstructed and moved while a concrete one cannot. The cities of the future might resemble a scene from the films of Hayao Miyazaki or the flying island of Laputa in Jonathan Swift's *Gulliver's Travels*; faced with a newly challenging environment, wooden cities could simply move elsewhere. Like snails, we would carry our homes on our backs.

In *Nomad Century*, Gaia Vince imagines how – in a world of runaway climate breakdown, where large parts of the Earth are uninhabitable – mobile timber cities might roam the Arctic like Howl's moving castle. Climate change is likely to force millions of people to migrate in the coming decades. Already, heatwaves in Asia threaten to make some of the world's most populous cities unliveable. But perhaps, rather than adapt our cities to a radically different and more challenging climate, we could re-engineer them to act as carbon sinks instead. Nature's greatest cities have been doing this for well over 500 million years. Coral reefs cover less than 1 per cent of the ocean floor, yet they support a quarter of marine life, drawing on the carbon dissolved in sea water to construct sprawling, intricate metropolises. These saltwater oases are threatened by ocean acidification, which eats into the calcium carbonate structures, and the rising temperatures that kill off the symbiotic zooxanthellae on which coral polyps rely for sustenance. Yet it may be that the model for the future city is not the forest island, but submerged in the waters that surround it. Could we *grow* the cities of the future under water?

Speculating that nature's solution to climate breakdown would be to grow more things from atmospheric carbon, architect Michael Pawlyn designed the Biorock Pavilion, a 3D-printed marquee that borrows the form of seashells. The entire structure is made of biorock or 'seacrete', an alternative to concrete and cement that is 'grown' in sea water: passing a low electric current through a submerged steel frame creates a calcium carbonate shell, which can be cast into any shape or size. Biorock's thermal and mechanical properties are very similar to those of concrete, it draws directly on the ocean's vast carbon and mineral reserves, and it can be produced on coastlines where nearly half the world's urban populations live. Imagine a city of similar buildings, all made with biorock: not only would this bring down carbon emissions, biorock can also 'self-heal', drawing down atmospheric carbon to mend cracks and fractures. Just like a reef, its porous surface would be an ideal substrate for many kinds of plant life, greening and cooling the urban heat island.

A few months after visiting Menno, I met with Michael in a café on Granary Square, in Camden, North London. It was a fresh and fine day, and as I was early I lingered in the large public space. Barefoot children darted gleefully between the jets of a fountain that rose directly from the paved surface. Moorhens bobbed placidly on the canal, while on the broad steps leading down to the water a team of workmen were laying carpets of plastic grass.

Michael arrived with a cap shielding his eyes against the bright March sun, and a loose scarf knotted against the coolness of early spring. We found a table in a quiet corner of the square, and he told me about how nature had inspired his approach to design. 'You have to really think things from first principles,' he explained. 'We should start by developing an understanding of the whole planetary system, and learn from that to rethink the elements we use, how we assemble those into materials, how we assemble those into bigger structures and also how we completely rethink the way we design cities.'

His ambition, he said, was to achieve a seamless integration of humans and nature.

Michael is a practitioner of biomimicry, a design practice devised by biologist Janine Benyus. Biomimicry is a way of making things that follows nature's lead. Before I met Michael, I had spoken to Janine over video call from her sunny kitchen in Montana.

She told me that biomimicry has three stages: nature as model, nature as measure and nature as mentor. Each represents a deepening of our engagement with what the natural world has to teach us. She gave the example of wind turbine blades modelled on the flippers of humpback whales. In 2008, inspired by observing a whale that had beached on the New Jersey shore, a biologist named Frank Fish devised a novel shape for wind turbine blades that mimicked the stippled surface of the whale's flippers. These bumps and hollows – called tubercles – allow water to flow smoothly over the flipper's surface by breaking up the eddies that form when flow reaches a certain velocity, giving the whales the ability to manoeuvre in tight circles. First of all, Janine explained, biomimicry is about moving from seeing nature as something to model, to seeing it as something to measure ourselves against. 'So we would go from a shallow biomimicry, which is "the scalloped edges of the humpback whale's fins reduce drag; let's put that on a wind turbine", to the next questions, which are "How are you going to make those scalloped edges? What system will it be a part of? Are you going to help with pollination or migration? And what will it nourish at the end of its life?"'

But the real challenge is to appoint nature as our mentor. 'Western industrial culture has not spent a lot of time learning from nature,' she said. 'Learning about nature, yes, but not *from* it. 'It takes a change of heart, not a change of technology alone.'

Michael's approach is similarly focused on a change of heart as well as a change in technology. 'If you can see something that works in biology, that is proof that it can be done,' he told me. 'But you could use biomimicry to just slightly improve lots of things, within a consumerist mindset. That's not going to deliver the transformation we need.'

Biomineralisation, the process of growing calcium carbonate structures, is only part of the process, he went on. His sharp eyes held my gaze. 'We should learn to rethink the elements we use, and how we assemble those into materials and bigger structures. There's the potential to integrate everything we do as humans into the web of life.'

Michael talked about designing with nature, not in terms of efficiency, but in terms of generosity. 'It's about trying to be as generous as biological organisms' – living in a way that 'maximises the potential of the system as a whole'. Building entirely with biorock would remove a major source of carbon emissions, and cool the urban heat island effect, easing pressure on those inhabitants who – unlike Menno's snails – are struggling to adapt. It could leave quarries and beaches to be rewilded; and at the end of a building's life, rather than go to landfill, it could be broken down and submerged as substrate for marine plants and animals.

Michael told me he was also working with a housing developer to make affordable housing from plywood and mycelium. The biological material 'eats its way into the wood, forming a very close bond, creating an incredibly lightweight, strong module with which you could build housing'. Mycelium conserves heat better than plastic-based insulation, and consumes waste as it grows. It could even be produced on-site, using plant fibres from the surrounding environment. 'And at the end of the building's life, you could just let it compost,' he explained.

I wondered how living in a 'seacrete city' or a compostable home might also change the people who lived in them. Winston Churchill famously observed that 'we shape our buildings; thereafter they shape us'. Yet they often do so covertly, quietly erecting ideological structures for our minds to inhabit. 'When the cathedrals you build are invisible,' writes Rebecca Solnit, 'made of perspectives and ideas, you forget you are inside them.'

The grandest, most encircling of these structures is the notion that we stand apart from nature. The city is perhaps the ultimate emblem of human achievement, an icon of our separateness from the rest of the living world. Working or living in a building made from biorock could dismantle the mental cathedrals that isolate and enclose us. Perhaps if we paused to listen, we would even be able to sense each sea-grown building quietly respiring – inhaling like N. K. Jemisin's living New York City.

Cities and homes grown in the manner of coral or using fungi could help us build new cathedrals of the mind. But changing our cities isn't just about changing their physical form. It's about changing underlying patterns; swapping the island mentality for the recognition that, whatever form they take, our cities are part of the whole planetary system and they need to work accordingly. Less islands, perhaps, than harbours, open to the wider world.

Michael invited me to walk with him to his next appointment, so we set off in the cold spring sunshine. On the way we passed an office block with a rippling glass front, as if a wave of progress were moving through the building. Yet it also gave an impression of weighty, impregnable solidity. Some people would argue that this kind of modern design is the pinnacle of human achievement, Michael observed, but it's almost universal practice for new buildings to have a 'non-infestation clause' – a contractual obligation to keep out pests such as mice or bugs, but effectively framed to encompass and exclude all non-human life. Future generations will look at these buildings as the embodiment of our separation from nature.

'What would the opposite of that be?' he wondered. 'How might we maximise inhabitation?'

Menno had asked a similar question during our winter walk several months before: as more wild organisms make their homes in our cities, 'how much space are we willing to concede?' The second lesson urban evolution teaches us is that cohabitation brings with it all kinds of exciting possibilities for doing things differently. If it's a truism that living together changes all parties, then it's as true for other urban species as it is of us. Ring-necked parakeets, which are endemic to India, arrived in London in the 1970s through the exotic pet trade (not, as the urban legend tells it, because Jimi Hendrix released them on Carnaby Street); by the early 2000s, when the parakeets were an established population, peregrines began nesting in the city and quickly learned to prey on them, and will even present parakeets as gifts during courtship rituals. In Seville, native birds have learned to nest near to parakeets, whose boisterous behaviour deters predators. New combinations lead to new relationships and sometimes new species; more than conglomerations of concrete and steel, cities are composed of relationships between living beings.

The same is true below ground. In 1968, 'sous les pavés, la plage' appeared graffitied on walls around Paris ('beneath the paving slabs, the beach'), expressing the desire and the belief that another city is not only possible, it actually exists, just beneath the surface of the one we inhabit. The metaphor has a literal dimension: every city possesses a unique microbiome, an assemblage of microorganisms specifically adapted to living in its soil, its sewers, on the porous surfaces of its buildings – even in its atmosphere – that won't be found anywhere else. Thinking about our relationship with the city beneath our feet could indeed help bring into being a livelier city on the surface.

Inhabitants of the urban microbiome are shaped by the food we eat and throw away, the waste we excrete and the antibiotics we consume. Many microbes like to group together to form biofilms, slimy multispecies communities within which various microorganisms coexist – much as diverse communities emerge in a modern city. When one of us wishes to settle in a new city, we make a series of choices and commitments: first, we choose the city; then we choose the neighbourhood; finally, we choose a home. The same process occurs when a bacterium becomes a part of a biofilm. First, it forms transient associations with other microbes on the surface. These new relationships make it easier to find the right place to settle down. Once this is identified, the bacterium can begin building its new home, adding to the depth and surface area of the biofilm. Just as in a modern city, for some life in the biofilm can begin to pale: for every new arrival who refreshes the community, another resident tires and resolves to try their luck elsewhere.

Biofilms, then, are in many respects like busy, bustling cities, composed of distinct neighbourhoods in which each tiny organism has a home. And if biofilms work like the communities that make up cities, can our homes and cities learn to be more like biofilms?

A few days after my conversation with Michael, I spoke to Rachel Armstrong, a professor of Regenerative Architecture at KU Leuven in Belgium, over video call. Her enthusiasm for designing with nature was infectious. By working with the metabolism of the urban microbiome, she explained, we can make our living spaces much livelier. Cities are politically important, but the tall, shiny buildings in the centre – whether they are built from conventional glass, or the siliceous spicules of glass sponges – are not where grassroots change happens. Rather than office blocks or public buildings, Rachel had begun thinking about how to redesign the home, and was drawn to microbes because they don't play entirely by the rules we've made for ourselves. 'Microbes are amazing!' she exclaimed. 'We grossly overestimate ourselves in relationship to microbes.' What they offer, she suggested, is 'a more rebellious notion of what it might mean to be human'.

'Big organisms like us have a preference for form,' she continued. 'That's not typical of microbes. They are highly organised, but they don't care for form like we do. We've dismissed them as being primitive because they don't like shapes like the ones we want.' In other words, we find the oozy formlessness of the microbial world unappealing. Who really wants to share their home with a slimy microbial community? Our generosity will only extend so far: the ick factor is a barrier. But we need to imagine spaces where our big, form-loving selves can coexist with what Rachel calls 'the squidgy, don't-care-about-form-but-like-relationships opportunism of microbes'.

Rachel's solution to this issue was to design a form for the microbes to inhabit, and she chose the most familiar – and perhaps even banal – shape she could imagine: the brick. 'Living bricks' are a combined microbial fuel cell and bioreactor, housed in stackable rectangular frames. They look, inoffensively, like white plastic milk crates, filled with vials of green water and crowned by a cluster of clear plastic pipes. Inside each brick, anaerobic bacteria create biofilms that, as they come into contact with energy in the form of household waste, both generate electricity and purify water. They are, she explained, a kind of 'interspecies trading system' in which humans and microbes mutually benefit. Panels of living bricks in a domestic kitchen or bathroom could recycle food waste or grey water (domestic waste water produced by washing machines and showers), while also powering the home. An organic economy of household waste would follow, trading waste with more nitrogen for waste with more phosphorus, for instance. Spare water and excess energy could be shared between neighbours depending on need. Each microbial community within a living brick would only involve microbes from the local environment – those that are unique to that city and are evolving in step with its more biologically complex residents via their food waste and chemicals. The result would be a home that not only evolves with the city but is, in a very real sense, alive.

Living bricks could radically change how we share our homes with other species – modest rectangles that could recast our relations with other beings in wholly new shapes. But they are also practical. 'It has to be manageable, conceivable, buildable,' Rachel said. 'That's why we chose the brick.'

The final essential element in the living bricks is data. To live with microbes, we need to be able to understand them. The biofilm in a living brick could be a communication medium, broadcasting whether the microbial community is healthy or running low on a particular chemical. Rachel showed me an animation she had commissioned to represent the constant stream of information coming from the living bricks. It looked like a cloud of butterflies.

'It gives you an overall view of how the biofilm is feeling,' she said. 'Because, you know, microbes don't have a very relatable face.'

So, like a kind of mood ring for the microbe? I asked.

'It is!' she exclaimed. 'It's a mood ring for the microbe, but it's digital. Or you can think of it as being a kind of pet, if you like.'

Then, after a pause: 'You know, we don't think of the water system or the electricity system in our homes as pets. But imagine if we thought of our homes as having a liveliness, a life and personality.'

Combining microbial gloop with rectilinear bricks won't make our cities carbon-neutral. But sharing our homes with living bricks would help us to see with greater clarity our place in a vast web of relations. Biofilms are egalitarian and cooperative by design: either the whole assemblage works together, collecting and sharing nutrients, or it fails. Our cities, by contrast, are riven by inequality (one study has even shown how racial inequalities between city neighbourhoods correlate with levels of biodiversity – the greater the inequality, the lower the biodiversity). Living bricks would draw us into what Rachel calls a 'commons of microbial processes'; that is, they would make our cities more like the urban microbiome.

To get to know a city, you need to meet it on foot. The philosopher Michel de Certeau once described Manhattan, seen from the 110th floor of the World Trade Center, as 'a wave of verticals', inhuman and abstract; the city from above becomes the city envisioned by planners and developers. Height can give us a synopsis, a summary of the urban narrative, but it is only at street level that we can become involved in the plot. To walk the city is to encounter it in all its recumbent possibility. On foot, we can step outside or around the official city – the walker is in a constant negotiation with the regimented spaces of urban officialdom, seeking short cuts, improvising new routes to get around obstacles, or simply open to the way any city can surprise us. As Rebecca Solnit writes, 'the magic of the street is the mingling of errand and epiphany'. Inspiration often arrives at walking pace; there is always the possibility that even the most unremarkable city walk will reveal something that transforms us or how we see the world.

This is the next lesson urban evolution can teach us about city life: to live well together, we need to improvise.

For Certeau, walking is like a form of speech; there is a 'rhetoric of walking' which speaks the city into being – where the metropolis organises possibilities and the walker translates them into actuality, writing the city with each step even as they read it. Every path taken composes 'the long poem of walking'. This is especially true of desire paths – the unofficial channels made when people cut corners and improvise new, unsanctioned routes from place to place. Some of these footfall-formed passages achieve a kind of rogue permanence, imprinting the urge to shorten a journey as an earth-brown line bisecting a patch of green; others are temporary, revealed by dark veins in snowfall. But if, as Certeau claims, cities are always in the process of being made by their inhabitants, then every passage through a city cuts a desire path; as the poets Paul Farley and Michael Symmons Roberts put it in *Edgelands*, their book of essays on the urban fringes, desire paths are 'proof of human unpredictability'.

Desire paths belong to a style of informal architecture called 'adhocism'. Coined in 1972 by architects Charles Jencks and Nathan Silver, adhocism describes 'a method of creation relying particularly on resources that are already at hand'; it is, they say, 'the style of eureka'. Adhocism remains a niche influence in human urban planning, but it represents the essence of how other species inhabit the city, improvising new means to familiar ends. Cavity-nesters such as sparrows and cliff-dwellers like falcons are especially able to find new uses for our buildings, transforming cracks, crevices, nooks and ledges into nesting sites. Since the 1980s, crows living in Japanese cities have understood how to use urban traffic to crack hard nuts. In a BBC documentary narrated by David Attenborough, they are shown dropping nuts in the path of fast-moving traffic, using the vehicles to gain access to the soft flesh inside the shells; the smartest crows have realised that, by positioning themselves by pedestrian crossings, they can also use the traffic of city walkers for protection while they collect the shattered pieces. By improvising new uses for what already exists, every urban animal composes its own desire path.

Towards the end of our walk in Leiden, Menno Schilthuizen took us past the city's central train station. In his book *Darwin Comes to Town*, he describes watching house sparrows flitting purposefully in and out of a two-storey open-air bicycle garage outside this station. Ranged together in close-fitting ranks, their handlebars and pedals intersecting and interlocking, the bicycle rack represents nothing so much as a large metal hedge, with the same thorny complexity as the birds' non-urban brushwood habitat. Protected by the tangle of spokes and saddles, they can forage in peace (for dropped crumbs rather than seeds and aphids), safe from predators. But it isn't the sparrows that have changed to fit the urban environment; instead, by using the city differently, they've altered it to fit their needs.

Other birds find shelter in the most unwelcoming forms imaginable. Anti-bird devices – arrays of steel spikes and thorns meant to keep messy wildlife away from the pristine facades of buildings – are a shameful urban innovation. But showing a laudable sense of irony, some birds have improvised nests from these mean-spirited devices. Carrion crows in Rotterdam and Eurasian magpies in Antwerp and Glasgow have been observed tearing strips of anti-bird spikes away from buildings to use as nest material. Mixed with more traditional twigs, the nests look jagged and uninviting – and one theory is that the anti-bird strips retain something of their original function, warding off competitors and predators. It is difficult to imagine finding ease in a nest that appears to offer all the comfort of a tangled ball of barbed wire, yet the birds occupy them contentedly, improvising shelter out of what was designed to exclude them. The spikes have other uses too: urban peregrines have been known to store leftovers of their prey on bird-control spikes, turning the devices into a kind of gruesome larder.

Biology doesn't plan, it improvises; the term for this in evolutionary biology is 'exaptation', coined by Stephen Jay Gould and Elizabeth Vrba to explain the origin of feathers. *Archaeopteryx* – one of the first dinosaurs to have feathers – could only fly very short distances. Flight wasn't the original function of the animal's feathers; *Archaeopteryx* couldn't fly any better than a pheasant. Instead, the feathers first evolved because those animals that had them were better at regulating their body temperature: they were adaptations to keep the creature warm, which over time were put to a wholly new use – exapted – to allow for flight. Exaptation places improvisation at the heart of evolution; the same process can be seen in the way some creatures make a home in urban landscapes, and learning from their talent for improvisation might show us how we can create lively cities in which all of life can be at home.

Simone Ferracina is an architect who helped Rachel Armstrong design the living bricks. He argues that we need a new kind of exaptive architecture if the cities of the future are to be sustainable; one that teaches us to pay a different kind of attention to what is around us – to stop seeing things only in terms of their original purpose, and start imagining their potential to be something new.

THE LIVING CITY

I met Simone at the College of Art in Edinburgh, where he teaches. Folding his tall frame into the seat opposite me, his soft voice sometimes hard to catch above the café's hum and chatter, he explained how we can reimagine our cities by drawing inspiration from the process of evolutionary change itself.

So many things are designed with a single purpose, he said, and with no thought to what might come after. The ultimate destination is landfill; the city is built astride a midden, to borrow from Samuel Beckett. A fixed idea of form – of what an object, a material or a building could be or do – can be a trap, Simone explained; but if we take the example of *Archaeopteryx*, or the birds building their nests from anti-bird spikes, then we can see that an object's original use is only the beginning. We need to alter the way we think about the time-scales of a lot of the built environment; we may design something for a particular purpose, but functions can change.

To many urban creatures, our cities are ready-made to be repurposed. Walking with Menno in Leiden, I had noticed a cluster of strange-looking trees lining the side of the road. As we got closer, I could see that they were in fact the chassis of scrapped and rusted cars, welded together and stood on end as arboreal-automobile sculptures. Jackdaws roosted in the twisted metal branches, just as if they were real trees.

Waste, for Simone, is all about potential. Waste isn't just what gets left over; it makes room for the unexpected to emerge. 'It's about beginning with something that already exists,' he explained – just as evolution works with what is available and then improves upon it. His favourite example was the Wikado playground in Rotterdam – a children's park made entirely from decommissioned wind turbine blades. 'Usually when we have objects that we don't know what to do with anymore, we either grind them into a lump we can reshape, or throw them away.

But these are very large objects, very difficult to move and impossible to grind.' The Dutch designers took the blades' pre-existing shape as their starting point, and played with the possibilities their form suggested. Simply by cutting a few holes and fusing parts together, they created a fantasy landscape of twisting slides and towers – inventing, in the context of a playground, novel uses and affordances. Blades that had once cut the sharp North Sea air became, at their tips, mountains to scale, and at their open bases, caverns to occupy; purely functional features were transformed into places where children could meet for a picnic or retreat and find solitude, or improvise their own sliding, clambering passage across the rolling surfaces. All it took to unlock new potential, Simone said, was a change in context. As he spoke, U2's 'Even Better Than the Real Thing' was thrumming over the café's speakers.

Wikado doesn't disguise the fact that it is made from objects that endured another life, out in the North Sea. Exapted features in living things often take this further, retaining their original role alongside the new function: feathers still insulate birds, as well as allowing them to fly; the first vertebrates evolved bones to store phosphates, and we still store nearly all our phosphorus in our bones. The park's sinuous forms summon memories of wild winter storms and the grace of diving whales, but their new life is one of play and possibility. 'For me, it's one of the best representations of exaptive design in the evolutionary sense,' Simone told me, leaning back in his chair.

Playfulness is an abiding part of a livelier city. Each of the projects I'd come across as I researched the lessons of urban evolution could have come from a child's imagination: What if we grew our cities underwater? What if we kept microbes as pets, and they also happened to power our homes? For urban creatures, the business of making a life in the city is a serious one, but there is also a sense of make-believe: What if this tower block was a nesting site? What if this cellar was a dark and welcoming cave? So it becomes.

A few weeks after speaking with Simone, I visited another project that could have come from a child's game. The Waste House is a two-storey house in Brighton, 90 per cent of which is constructed from things other people have thrown away. More than 8 million tons of waste goes into landfill in the UK every year; in the US, it's nearly 140 million. Globally, we produce over 5 million tons of waste *every day*. But we don't see it. In Don DeLillo's novel *Underworld*, a 'garbage archaeologist' and junk visionary called Jesse Detwiler expounds on the treasure to be found in trash. 'Bring garbage into the open,' he declares, standing on the lip of a vast engineered crater that is destined to become landfill. 'Let people see it and respect it . . . Make an architecture of waste.' The Waste House seeks to make this prophecy a reality.

The Waste House looks a bit like the cottage that enchanted Hansel and Gretel, if it had been made from junk instead of sweets. On the day I met the house's architect, Duncan Baker-Brown, an overcast raw-looking sky loured blackly above us and his bright yellow scarf glowed against the Waste House's dark exterior. The first thing he invited me to do was touch the outside walls. They were unexpectedly rubbery: from a distance they had looked like wood panels but in fact the whole house was clad in reused plastic carpet tiles, lapped like slates with their rubber undersides facing out. On some, the printed catalogue numbers were still legible. Slipping my hand beneath the tiles, I could feel the felted tops. Duncan told me they had been recovered when the next-door building was renovated. Despite being designed for use indoors, they were 99 per cent rainproof.

'The project was initially designed to prove that construction waste is valuable,' he said. Twenty per cent of the UK's waste is produced by the construction industry. But then he began to think about how he might make use of everyday waste as well. The roof was covered in repurposed tyres. He tapped a concrete paving slab with a toe. 'These were reused two or three times before they became this path,' he told me with pride.

We sought shelter inside as it began to drizzle. A warm, woody smell greeted us, and I was struck by how homely this house made of trash felt. In the single downstairs room, post-box-sized windows revealed the cavity between the inner and exterior walls. One appeared to be filled with a jumble of toothbrushes; in another, DVDs were lined up as if in a library, their spines showing the faces of Disney's Mowgli, Molly Ringwald in her 1980s prime, Al Pacino as Michael Corleone, and a grimacing Marge Simpson. I guessed that the partially obscured title —*g Day* was probably *Groundhog Day*, a film in which the same events repeat in an infinite loop.

Duncan described the Waste House as a lesson in how and how not to do things. The household items they'd used as cavity insulation – not just toothbrushes and DVDs, but also duvets, floppy disks, rolls of wallpaper and bicycle inner tubes – were a poor substitute for natural alternatives like sheep's wool. There were 25,000 travel-sized toothbrushes in the walls, he said, all of them collected in just four days from Virgin Airlines, most of them unused. The reason for putting them there was to show people that we need to 'reimagine waste' altogether. The way the Waste House had been invented from the most unlikely materials was delightful, but it was evident too that, while some improvisations – like rainproofing the outside walls with the carpet tiles – were inspired, others were a less good fit for their new purpose.

We climbed a staircase made of paper up to a pleasant, Scandinavian-style loft. Despite being composed almost entirely of junk, the Waste House is safe to live in. The university regularly uses it to hold workshops, and even hosts tours of the house for local schoolchildren. The loft had a cosy, inviting feel, heightened by the light rain that had begun to speckle the windowpanes. But not everything in this warm, comfortable space looked hospitable. Duncan pointed to the lights suspended from the asymmetrically pitched roof. They were made of steel and looked heavy and industrial. He explained they were eighty-year-old blastproof light fittings, stripped from a decommissioned South Korean container ship at Chittagong, a port city in Bangladesh where old ships go to die and be stripped for parts. 'A child recovered these, probably working in toxic conditions. But I bought them from a dealer in Italy.' He paused. 'It's a question I ask every visitor: should *they* be here?'

'Buildings are accretions of labour, energy, carbon, and environmental damage,' Simone Ferracina had told me. The light fittings, hanging weightily on their chains, were heavy with a history of exploitation that hadn't evaporated when they passed through the Italian dealer's hands. Every material we currently build with is, similarly, weighted by a history of exploited people and natural resources, often wrapped up in colonial legacies – whether in the displacement of indigenous communities to gain access to minerals or fossil fuels, or the implicit designation of the global south as a 'sacrifice zone', set aside to receive the rich world's discards and exposed to the worst effects of climate breakdown. If, as Simone insisted, waste can speak to us of possibility, it should also continue to speak of its past. Another benefit of an architecture of waste is that it shows us our place in the circle of complicity.

The Waste House's junkyard Hansel and Gretel aesthetic

discloses a deeply serious truth about the kind of play that turns a wind turbine into a playground or part of a container ship into a designer light fitting. Such improvisations might save the materials from landfill, but whatever new use we discover in them, some damage has already been done. Moreover, much of what we throw away is locked into a particular form, and unlocking it to release the potential to become something else can be difficult and costly. Duncan spoke about how we might begin to mine materials like aluminium from what we throw away. Janine Benyus had expressed a similar idea. 'Yes, we're gonna have to start mining landfills,' she said; but ultimately, she insisted, we need to start designing materials with an eye to reusing them. The things we make, from concrete to plastic to textiles, are embodied energy, and limiting what we can do with them represents an improvidence that has no place in nature.

Where Duncan and Simone imagine a future city as a kind of shipwreck, a bricolage of elements and parts broken and repurposed, Janine prefers to think the next age of urban living will involve more of a return to nature. 'When you look at a healthy forest and you look at a stream coming out of it, there are very few nutrients in that stream,' she said. 'I mean, they're not leaky like our cities. The reason they're not leaky is they've got all those systems of reuse and transformation and decomposition. When something dies and falls to the forest floor, the material gets picked up into the body of another organism. Some of it is pooped out as faeces, but most of it is metabolised by that body and then another organism eats that one.' She gestured left and right. 'And then the material goes here, or goes here.'

On the Biomimicry 3.8 website, Janine tells a story about what the forest city of the future might look like. The year is 2050 and you are approaching the city on a translucent jet, gliding silently over a landscape carpeted by a vast forest. When it appears, the city is difficult to distinguish from the forested wilderness surrounding it. Just as 'a forest creates goodness and then gifts it away', she writes, 'the city is as generous as the wildland next door'. Benevolence, modelled on the local ecosystem, has been designed into its fabric and infrastructure: for instance, the entire urban edifice is a vast water-cycling system, gathering rainwater via absorbent pavements and restoring it to aquifers; roof canopies are designed with undulating surfaces to increase evaporation, restoring water to the clouds. Corridors of agricultural land and passages for migratory animals are threaded through the streets. It is 'a city in a forest and a forest in a city'.

Real-life versions of Janine's hyper-modern forest city already exist. Lavasa is a planned eco-city in the Pune district of western India, a bit more than 100 miles south of Mumbai and around one-fifth the megacity's size. Built from the ground up by a private company on what was intact broadleaf forest, it illustrates both the potential and the cost of Janine's image of a generous city. Following Biomimicry 3.8's vision, Lavasa has roads modelled on anthills, which remain in place despite the heavy monsoon rains by directing water through a series of sinuous channels that slow the flood; the foundations of its buildings mimic the roots of trees, and its pavements and roofs capture and release water back into the ground and air. But several untouched sacred groves were dismantled to make way for the city, which is also serviced by a six-lane highway shuttling residents between Pune and Mumbai – the construction of which displaced sixty-six villages and nearly 100 hectares of forest. The road is the only way in or out of Lavasa.

Janine's forest-city vision is immensely appealing, but the reality of Lavasa suggests that we can't presume to build our way to a sustainable future. There are no tabulae rasae on which we can erect new cities to replace the old; we need to improvise alternative versions of the cities we have. Rethinking what waste is could help birth a circular economy, taking us out of the damaging cycles of extraction and consumption in which we're trapped. Simone stressed how we need to leave space open in order for those who follow us to reimagine the materials we use – to discover what is possible with these materials that we can't yet see. But true circularity will only be possible if we learn to design flexibility, reuse and (eventually) decomposition into everything we make. Perhaps, in the end, the cities of the future will need to be a bit like a forest *and* a bit like a shipwreck, drawing a spirit of generosity from one and a spirit of improvisation from the other. What must be built new should cleave to the forest; where something can be repurposed, then we should act like castaways – like Robinson Crusoe, making a new life out of the wreckage. But in the end, generosity and improvisation both begin with an imagination that is open to what could be.

As I left the Waste House, I thought again about the birds who fashion their homes from anti-bird devices. Dressing our buildings in spikes to keep birds from roosting indulges our most miserly impulses: hoarding space that ought to be shared. It takes a generous imagination to see how a barrier can become a shelter.

Somewhere out in the dusk a blackbird was singing; a small, bright point in the darkening city.

In 2017, two musicians – Tim Vincent-Smith and Leon Wright – decided to build a performance space out of salvaged pianos. Pianodrome consists of three wedge-shaped stands arranged in a circle, each one constructed entirely from cast-iron harps, side arms, legs and assorted other piano parts – even recovered nails. Audiences can climb soundboard staircases with curved banisters made from reclaimed piano lids, and rest their heads against seatbacks made from old keyboards. A working upright piano is built into each stand. When I went to visit Pianodrome, it was being kept in an empty department store in the north of Edinburgh.

Slim, with brown hair tied in a loose ponytail, Tim was playing a grand piano at the store entrance when I arrived. Instead of rails of clothes, the space was filled with row upon row of old pianos, as well as sculptures made from old instruments. Tim and Leon were running an 'adopt a piano' scheme for anyone who couldn't afford an instrument. 'We say that no piano is junk and no person is unmusical,' Tim told me. 'Everybody is musical, everybody is creative. It's just a question of making spaces where that is accepted and understood.'

Pianodrome itself was behind an eerily still escalator. The old store was filled with the sound of people playing, some hesitantly, others confidently, melodies clashing or winding round one another. We took mugs of tea and sat in the Pianodrome space. An old piano stool cushion covered the seat next to us, embroidered with the message 'The world is yours'. Tim balanced his mug in the crook of his elbow as he spoke, sometimes having to raise his voice above the echoing cacophony.

'Even if they aren't useful as playable instruments, there's still so much beauty and so much useful material in them,' he said. 'We want to suggest that can be done with people as well. Who cares if your music teacher says you're not musical, or the government cuts all the provision for music: this is an inherent human activity!'

Pianodrome was constructed only of instruments thought to be beyond repair, extending their life of making music. It unlocked something hidden in what was discarded. Broken, unwanted things were given a new lease of life. But it also unlocked something in the old department store. As well as the piano sculptures, children's art covered the walls. Above them, a sign had been altered to read: 'Please p/ay here'. This space of commerce had been transformed into one of imagination and fun. It seemed like a glimpse of what a better city could look like.

If Pianodrome were scaled up, everything in the city would retain the potential to be otherwise. Just as evolution never rests, the city would never be considered finished or closed off. Like the birds building nests from anti-bird spikes, what was built to exclude could offer shelter; as crows use passing traffic to break open their nuts, the paths we would compose through the city could also nourish others. Nothing and no one would be discarded as waste. Every structure and surface would tell a story: of what it was, what it is now, and most importantly, what it could be.

The store was silent. But when I pressed a key on one of the working pianos, the space filled with sound. The carpenters had left the strings on the old piano harps in place, buried inside each wedge of the amphitheatre. They resonated with and amplified any sound made inside the stage. The whole thing, I realised, was a giant soundboard. It was as if, with each new note played, the discarded pianos were recalling their old lives, even as they hymned the new.

I asked Tim what it was like to play here. 'The acoustics are *amazing*,' he said emphatically. 'The quieter you play, the more powerful it is, somehow. I've found that, as a musician, the most moving experiences have been when it's been so quiet, you're almost imagining the sound.'

Sometimes you can hear it, writes N. K. Jemisin, faint beneath the urban clamour, but there: the living city, breathing.

THE LIVING CITY

For now, Pianodrome's future was uncertain. The old department store was due to be demolished, and it wasn't clear where they would take it next. But Tim was undaunted. 'We take opportunities where we can,' he said, smiling, 'like a seed in a crack in a wall.'

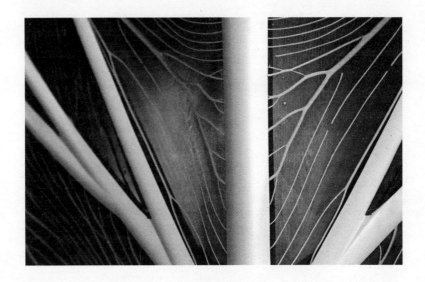

Aguahoja III (detail)

3

ONE TOUCH MAKES THE WHOLE WORLD KIN

How nature can help us fix our waste problem

In Olga Tokarczuk's novel *Flights*, the narrator observes the appearance of a crowd of strange new creatures. Part animal, part plant, they travel in great flocks, like birds or the seeds of wind-pollinated flowers. 'I see them from the window of the bus,' she notes, 'these airborne anemones, whole packs of them, roaming the desert.'

Their emergence is rapid and apparently inevitable. Already, the narrator tells us, they have colonised every continent and almost every ecological niche, from highways to the slopes of the Himalayas. Borne by the wind, distance does not trouble them. Regardless of where they alight, they use their large, prehensile ears to attach to whatever they find – rocks, trees, fences, telegraph poles, even other creatures. At first glance, the creatures seem frail, shredding as easily as the gossamer bodies of butterflies. But they prove surprisingly durable.

The strange animals are, of course, plastic bags. True to their name (with its root in the Greek *plastikos*, 'mouldable'), they have achieved 'pure form', an innovation that 'affords them great evolutionary benefits'. Hollow and composed entirely of surfaces, they insatiably seek out 'contents' to fill the void within. Yet whatever they encounter, they are themselves never changed. For all its malleability, plastic is always resolutely itself; whatever form it moulds to, its long-chained molecular structure does not alter. Under stress, it simply breaks down into smaller and smaller versions of itself. Plastic's close-knit polymer bonds are what give it such a lengthy life, but its plasticity is the gift of a host of chemical additives, which can account for up to half the mass of the final product. As they are not a part of the polymer bond, these toxic substances can very easily leach into the environs. Every one of the between 15 and 51 trillion microplastic particles drifting in the world's oceans delivers a quotient of flame retardants, surfactants, plasticisers and dyes into the marine environment.

As in Tokarczuk's fable, plastic seems to have attained ubiquity today. It is present worldwide in soil, ocean water and sediment, and also Arctic ice. It falls as rain, invading the phloem of living plants, and even passes through the placenta to enter the bloodstream of babies in the womb. It is embedded at the very base of the food chain. But plastic is just one facet of the chemical age. According to a 2017 report, 10 million new chemical compounds are synthesised every year, at a rate of more than 1,000 per hour. Many find their way into the world without us really understanding their effects.

Vast quantities of nitrogen and phosphorus washing off American farms and into the Mississippi River watershed have created more than 6,000 square miles of hypoxic water in the Gulf of Mexico – an oxygen-free dead zone in which only jellyfish and microbial life can survive. Pesticides are wiping out vital pollinators, with dire implications for many bird species. Heavy metals from mining and industry salt the earth and foul the air, often far from where they were extracted. Methylmercury, a by-product of gold-mining in the Amazon, burns in the bodies of Arctic wildlife. Lead, cadmium, nickel, beryllium and zinc wash up in developing countries as e-waste, weeping from discarded laptops and mobile phones.

Not all life toils under this toxic burden, however.

'A toxin threatens,' writes the philosopher Mel Chen, 'but it also beckons.' A select number of plants and animals have responded to this summons, developing the remarkable capacity to thrive in environments so polluted that nothing ought to survive in them. The Juan Fernández Islands, where Alexander Selkirk, the inspiration for Daniel Defoe's *Robinson Crusoe*, was marooned in 1704, was once home to a population of around 80,000 fur seals. It was thought that fur trappers drove the species to extinction, until a small population was discovered in the 1960s. This remnant community can tolerate extraordinarily high levels of heavy metals. The seals travel far into the vast, churning sump of the Pacific gyre in search of fish and squid, where they also consume polymers saturated in toxins like cadmium. Scientists don't yet know how the animals are able to remain unaffected by a substance which, in humans, can cause cancer and pulmonary disease.

Some organisms adapt to toxic environments extremely swiftly. Water fleas living in a German lake were discovered to have developed a tolerance for toxic cyanobacteria – a consequence of eutrophication, the process of oxygen depletion that produces dead zones – within ten years. Analysis of the genes of insects that died in the 1960s and were recovered from soil sediments shows that they wouldn't have been able to cope with the deadly algae, but those born in the 1970s could. By the turn of the millennium, the entire population was able to abide cyanobacteria.

For some, pollution even represents an opportunity. One

species of frog found that taking a chemical bath offered a remedy to chronic infection. The rapid global spread of a deadly chytrid fungus, *Batrachochytrium dendrobatidis* (or *Bd*), has led to the extinction of around ninety frog species since the 1970s, along with serious declines in around another 500 amphibian species; a quarter have lost more than 90 per cent of their population. The fungus causes the animals' skin to harden and slough off, preventing them from taking in fluids. *Bd* emerged on the Korean Peninsula sometime in the 1950s and has since followed international trade routes to Australia, North, South and Central America, the Caribbean, and the Iberian Peninsula.

In 2013, thousands of green and golden bell frogs emerged from a water treatment facility in Homebush, a suburb of Sydney, Australia, having taken a chytrid-cleansing bath in the swill of domestic waste. Homebush was once the site of the largest abattoir in the Commonwealth. A vast early twentieth-century open-pit brickworks in the area later became a forty-five-foot-deep garbage dump. The site was remediated for the 2000 Summer Olympic Games, with the old brick factory transformed into the sewage works. Some studies have suggested that ordinary domestic pollutants such as soaps and laundry detergents, as well as the antibiotics that are present in wastewater, washed the frogs clean of the chytrid fungus. They seemed content in their bath of household chemicals, though observers noted that every frog that emerged from the leachate ponds was a vivid lime green.

Other species adapt their appearance, specifically their colour, to cope with polluted habitats. Urban pigeons with darker feathers are better able to handle the potentially harmful substances that are a part of city life, because toxins like zinc bond with melanin in their feathers rather than corrupting their bodies. It's a trick also learned by sea snakes in New Caledonia.

They typically wear a combination of light and dark stripes, but those that live around the capital, Nouméa, where a metallurgical plant pours its waste into the waters, have evolved to fill in the gaps between the darker stripes. In fact, industrial melanism, as the phenomenon is sometimes called, is the earliest instance of an adaptation driven by human influence. It was first observed in the middle of the nineteenth century, not in birds or reptiles, but in moths.

In the 1840s, in the woods around industrial Manchester, a moth that was once snowy white appeared to have turned dark as coal. The first specimen was caught and pinned in 1848; by the 1860s, the black-winged variant was more common than the original white. By the end of the century, the pale peppered moth was all but eliminated. A 2016 paper in *Nature* traced the transformation to a mutation that had occurred in a single caterpillar in 1819: a lone transpon, a so-called jumping gene that controlled wing pigmentation, had travelled from one chromosome to another. The new grimy colouration allowed the moths to blend in with the polluted landscape (acid rain had killed pale lichens, and soot coated the branches of surviving trees); birds would pick off the lighter moths, which shone brightly against the blackened trunks.

'Evolution favours fleeting forms,' as the narrator in Tokarczuk's *Flights* notes. The story of the peppered moth didn't end with a black insect camouflaged against a blackened tree. Between 1965 and 2005, at nearly the same rate as the black-winged population had increased, pale moths began to supplant them. The 1956 Clean Air Act was scrubbing away much of the residue of Britain's industrial past, and by the turn of the century, the ratio had reversed, with barely any black moths left.

But 1965 was also the year another new form began to take

shape, one that would come to dominate the landscape even more effectively than the peppered moth. That same year, a Swedish designer named Sten Gustaf Thulin applied to patent a new kind of sack for household use, exploiting the possibilities of a novel, lightweight, highly malleable material; the plastic bag took flight.

What does it mean to inhabit a world so polluted that our traces, reaching deep into the bodies of other beings, are moulding those bodies into new forms? Our chemical world is beset by hidden harms. Injury lurks in air, soil and water. But toxins that threaten also beckon to us, inviting us to contemplate what ties together individuals and industries, communities and chemicals, across space and time. They point us towards the latent potential to be otherwise that lies in every cell of every organ of every living thing. As strange as it may sound, pollution gives us a window onto a world in which change is possible.

The first sense we develop in the womb is the sense of touch. At eight weeks, when it is little more than a spermaceti squiggle with eyes like black pips, a foetus is alive to the world. Touch receptors cluster around the lips and nose. At twelve weeks, receptors bud in the genitals, and on the palms of the hands and the soles of the feet. Twins in the womb will actively seek one another out, preferring to caress their wombmate than touch themselves. By thirty-two weeks, the foetus's whole body is so sensitive it can detect the pressure of a single hair.

In 2021, when much of the world was deprived of contact by the Covid-19 virus, the Nobel Prize in Physiology or Medicine went to the discovery of the receptors that recognise temperature and touch. We know the feel of the world around us because of pressure-sensitive cells that can respond to a protein called Piezo1 (from the Greek for 'pressure'); another protein, Piezo2, is critical to proprioception, the awareness of movement and the body's position in space. Without these receptors, we would not know our own hand as it moved past our face, or our foot as it touched the ground. The same pressure-sensitive cells tell us when our bladder is full, and how much air is in our lungs.

Scientists are only just beginning to grasp how important touch is in shaping who we are. Its absence is catastrophic for the cognitive and emotional development of infants. A series of studies by psychologists at Yale and Harvard found that handling different weights and textures directly influenced abstract thought. 'Everything we love or lose,' writes Fernando Pessoa in *The Book of Disquiet*, 'brushes our skin and thus reaches our soul.' Touch composes us: we metabolise it, drawing it into ourselves. We are composed of a lifetime of touches.

But what of that which touches us without our knowledge? Because of Piezo1 we can sense how much air fills our lungs, but not the contaminants and microplastics mixed with it. In this chemical world, toxins trespass the limbic system incognito; pollutants infiltrate our cells unnoticed as well as unbidden.

Over the course of seven years, the Canadian poet Adam Dickinson submitted his blood, shit and urine for toxicological analysis, in order to discover which industrial chemicals he had ingested and absorbed from his environment. The substances he was tested for included phthalates and bisphenol-A (BPA), absorbed from everyday plastics, as well as polychlorinated biphenyls (PCBs), perfluorinated chemicals (PFCs), pesticides, insecticides, parabens, flame retardants and thirty-one heavy metals. In total, he filled seventy-eight vials of blood. Initially wary of needles, by the end, he writes, 'I was drawing from both arms and yanking on the tourniquet with my teeth.'

Anatomic, the book of poems that came from his investigations, describes the troubling discoveries he made. PCBs marbled his fatty tissue; phthalates crowded his endocrine system. Uranium from groundwater and from nuclear testing was embedded in his bones and teeth. 'I am a spectacular and horrifying crowd,' he observes. 'How can I read me? How can I write me?'

Over a scratchy video connection, Adam explained to me where the process began. He had been writing poems about plastic pollution, and became curious about where it touched his own life. 'We're rewritten by our chemicals. I became interested in thinking about myself as a site of metabolic writing.' Military, industrial and agricultural histories had accumulated in his tissues, including PCBs manufactured by Monsanto, and DDT, a once commonly used pesticide that was banned before he was born.

Many of the substances Adam found in his body had the potential to interfere with his endocrine system, the network of glands that controls hormonal traffic. Endocrine-disrupting chemicals, or EDCs, affect the ecology of gene expression, which is to say they trigger changes in gene function without causing a mutation. Instead, they allow possibilities latent in the gene to become reality. Every one of an organism's cells contains its entire genetic code: for a cell to perform the correct function – to become a brain cell rather than a cell buried deep in the muscle of the heart – certain genes within the cell must light up, while others are kept dark. Hormones are activated in the endocrine system when they find a receptor whose shape they fit, like a hand slipping inside a glove, which nudges the cell to make a protein or express genes that influence how the body metabolises or grows.

As of 2013, the World Health Organization had identified

more than a thousand chemicals with endocrine-disrupting properties. Some are in insecticides, herbicides and fungicides, but many are used to manufacture domestic products such as plastics, cosmetics and toiletries. EDCs mimic the shape of hormones like oestrogen, fitting as neatly with the receptor as natural hormones do, but in doing so they edit the function of the cell, warping the story it tells – and many are far more potent than the hormones found in nature. Since the time of the Renaissance physician Paracelsus, the primary toxicological maxim has been 'the dose makes the poison', but EDCs violate this principle. Their effects are not dose-dependent, and can be more powerful at lower doses. Infants and foetuses, whose still-developing bodies are the most plastic, are most vulnerable. There is, in effect, no safe threshold. And the touch of some EDCs is extraordinarily delicate. Just a single molecule of synthesised oestrogen, attached to the right strand of DNA, can rewrite the phenotype, causing cells to express genes differently or to manufacture an altered kind of protein, and potentially leading to cancers, infertility or developmental disorders. Endocrine-disrupting chemicals slip anonymously into receptors like agent provocateurs, gaining the body's trust before wreaking havoc.

Adam's poems read like echoes of the chemical poems being written in his flesh – ghostly imprints of a composition taking place deep within his tissue. Sometimes the effects are eerie. 'Try to place the chemically manufactured smell of fruits and flowers,' he notes in a prose-poem called 'Mono-isononyl phthalate', named for a chemical found in his urine which is used to make scented candles, 'and you find yourself on a debris flow into the uncanny valley . . . *Pumpkin Spice* is like hugging your grandmother, inhaling deeply, only to feel some of the stuffing emerge from the crease in her neck.'

The effects of EDCs on wildlife were first noticed in the 1980s. Inspired by an experiment which demonstrated the sensitivity of foetal mice to uterine hormones (female mice positioned together in the womb were exposed to more oestrogen, and were less aggressive in adulthood than those exposed to androgens via their brothers), environmental health analyst Theo Colborn wondered what effect chemicals polluting the Great Lakes were having on wildlife. The Great Lakes region is home to a fifth of American and half of Canadian industries, many of which discharged their waste indiscriminately into the region's rivers and streams for decades. Colborn discovered that fish swimming in this noxious soup had altered sexual characteristics. Male carp and walleyes living near the outflow of sewage plants were producing vitellogenin – an egg-yolk protein made by females – rather than sperm, and male white perch were developing intersex features. Round goby – a bottom-feeding fish whose mouth hangs open in a perpetually startled expression, introduced to the region in the ballast water of commercial ships – exhibited the full range of effects, including a skewed sex ratio, shortened male urogenital papilla and the production of vitellogenin by male fish.

Similar phenomena were observed elsewhere in the US, and beyond. In Florida, female mosquitofish exposed to testosterone-mimicking paper mill waste developed gonopodia – anal fins that male fish use for copulation – while male alligators' penises shrank by up to half their normal size. On the West Coast, male Chinook salmon developed female characteristics. In Svalbard, polar bears were being born intersex.

The realisation that endocrine disruptors, so pervasively present in the environment, could alter the sex of an organism led to a well-publicised media panic. Hysterical headlines played on masculine anxieties in particular. 'Pesticides "Castrate" Male Frogs', they howled; 'Otters' Penises are Shrinking – And Why Yours Might Be Too'. Synthetic oestrogen was first produced commercially (and thus entered the environment) in the 1940s, and since then sperm counts in the US have been cut in half. Addressing the House Subcommittee on Health and the Environment in 1995, Louis Guillette, the biologist who discovered the decline in alligator endowments, made an announcement guaranteed to send a shiver down the spine of every red-blooded American man. 'Every man in this room today is half the man his grandfather was,' he intoned ominously. 'Are our children going to be half the men we are?'

This association between pollution and emasculation not only reinforces prejudice, it fails to understand the way sex emerges in the body. There are countless examples in nature that demonstrate biological sex is not a fixed state but rather a matter of potential. Temperature, the pH of water, even social factors can all alter the sex ratio of a group. Some beings blithely change sex many times during their lifetime. Marine snails are male when young and transition to female as they age. Coral goby switch between male and female depending on their environment or the availability of mates, as do black sea bass. Others mimic the opposite sex: female *Papilio phorcas* butterflies can clothe themselves in the apple-green brilliance of male wing patterns to avoid predators, and male mason wasps, which have no venom, brandish genital spines that resemble a female wasp's sting. For much of the natural world, binary sex is an unnecessary limitation. Most plants are intersex; to ensure total genetic diversity within its breeding pool, *Schizophyllum* fungi have upward of 28,000 different sexes.

The carnival intensifies at a smaller scale. As well as toxins,

Adam Dickinson tested his microbiome, the ecosystem of bacterial life that supports the immune system, aids digestion and even regulates mood. 'My body is a spaceship designed to optimize the proliferation and growth of its microbial cosmonauts,' he writes in *Anatomic*. These bacterial passengers are truly alien to our concept of what it means to be a living thing. Exploding any notion even of species – never mind sexual – difference, bacteria engage in a form of permanent symbiosis, merrily swapping genes with any organism they meet. Touch leads to transformation. The bacteria in our bodies, notes sociologist Myra Hird, are 'infinitely sexually diverse'. Binary sex is a fairly recent evolutionary development. For most of the history of life on Earth, sexual difference has been a festival of variation and possibility.

The cells of my microbiome outnumber the human cells in my body by between 1 and 10 per cent. More of me is beyond sexual difference than defined by it. Moreover, every cell in my body has the potential to change its function, for good or ill: metastasising into grotesque tumours, or animating a new potential in my genome.

Endocrine-disrupting chemicals can do terrible harm, but they also beckon us to think again about the plastic potential of any living organism, ourselves included. Nor should we be quick to condemn the substances themselves. Hormones – even synthetic ones – are morally neutral; their affordances depend on their context. For trans philosopher Paul Preciado, testosterone is 'an ally . . . triggering a parallel evolution of my own life, by giving free expression to the phenotype that would otherwise have remained silent'. Really, all an endocrine disruptor does is flick a switch in the genome; contact with these vivid chemicals can touch off a kind of evolution – another self buried deep within.

ONE TOUCH MAKES THE WHOLE WORLD KIN

Adam's exploration of the chemicals saturating his body shows there is a poem being made in every one of us, written by many different hands. It can be alarming to realise we are not the final authors of ourselves, especially in a chemical world where harms are ever-present; but it can also be enticing. A toxin can beckon us to recognise that we are more than we think we might be.

'Where is the beginning and the end of touch?' Adam said towards the end of our conversation. 'The things we touch also touch us back.'

And of that which touches us, what sticks, what stays?

From the eastern shore of the Hudson River, the glittering towers of downtown Manhattan face their counterparts in Jersey City across the wide mouth of the estuary. Even on an overcast day, the tallest and grandest have a fish-scale iridescence, multiplying the city in thousands of panes of glass.

Between the river and the flow of traffic on Route 9A, Hudson River Park edges the western shore of Manhattan. On a map, it looks like a fringe of moss on rock. For joggers and strollers, it offers respite from the bustle and noise of the city's deep canyons. It is also a refuge for monarch butterflies, which use the park as a waystation on their epic, 3,000-mile southern migration to Mexico. It was too cold for the butterflies when I was there in late September, but a few weeks earlier the park had worn a flickering coat of amber as numberless migrating monarchs descended on Manhattan.

My interest was not in the park, however, but the river. The Hudson emerges in the Adirondack Mountains and follows a 315-mile route through New York State and into the Atlantic Ocean. Presently named after an English explorer, it was called Muhheakunnuk by the Lenape people (meaning 'great waters constantly in motion', reflecting the fact that the river is tidal for around half its length). Tremendous wealth flowed down the river and into the city in the nineteenth century, with the opening of canals connecting the Hudson, and thus the Atlantic, with Lake Erie. By the next century the flow had become distinctly brackish, as wealth mixed, like salt and fresh water, with vast quantities of industrial waste. It is said that, by the 1970s, it was possible to know what colour General Motors were painting their cars by the colour of the water in the lower Hudson.

Heavy metals and petrochemicals poured unchecked into the river for decades, but by far the worst pollutants were polychlorinated biphenyls (PCBs), one of a class of industrial chemicals known as persistent organic pollutants – or, more colloquially, 'forever chemicals' – for the extreme slowness with which they break down. For almost thirty years, between 1947 and 1976, General Electric released 590,000 kilograms of PCBs into the Hudson, from factories located in Fort Edward and Hudson Falls. The contamination was so great that, in 2002, the Environmental Protection Agency designated 200 miles of the Hudson River a Superfund Site, a status reserved for the most contaminated environments and thus a national priority for remediation. Indeed, the Hudson is the largest Superfund Site in the United States. So far, $7 billion has been spent on cleaning it up, but the PCBs remain. Overall, around 1.3 million tons of PCBs have been produced worldwide – added to paints, caulks, plastics and domestic detergents. At least a third have made their way into coastal sediments and the oceans. Highly hydrophobic and lipophilic (which means they dissolve more readily in fats than water), they can lodge in the fatty tissues of anything living. They are probably present in every person alive today.

ONE TOUCH MAKES THE WHOLE WORLD KIN

PCBs can be highly toxic, but one species of fish in the Hudson has evolved the ability to shrug them off like a bad mood. I was in the United States to do some teaching at Rutgers University, across the river in New Jersey, and had a day or so in New York. I had taken a subway train from Midtown to Hudson Yards, and was walking south through the park to the Hudson River Park's Pier 40 Wetlab to find out about this remarkable and resilient species of fish: the Atlantic tomcod.

The southern side of the pier was mostly empty when I arrived, with just a few people fishing off the wharf. On a bench beneath the huge pier shed, a young man was shaving off his friend's dreadlocks. On the side of the shed, painted in joyful red letters several storeys tall, a mural memorialising the AIDS pandemic proclaimed 'I WANT TO THANK YOU'.

A bubble of noise rose as I arrived at the entrance to the Wetlab: a party of schoolchildren were visiting to learn about the different species of fish and crustaceans that live in the river. Navigating the cross-currents of excited children, I made myself known to the lab staff, then stepped outside to wait. It was almost as loud outside as it was inside, with the whump of helicopters browsing the sky above the Hudson. As I waited a tall figure with a grey beard and a khaki jacket approached along the pier. I guessed he was in his late fifties or early sixties. Thick glasses magnified soft brown eyes.

Since the 1980s, Ike Wirgin has researched the effects of pollution on tomcod in the Hudson River. Fish have always been his thing. Growing up in Mount Vernon, New York, he was entranced by the tanks in the local fish-store window. Later he began fishing, but more for the chance to study his catches than to eat them. 'In the early 1980s, there was a lot of concern about reports of tumours in fishes across the

United States,' he told me. 'There were tumours in winter flounder from Boston Harbor, in brown bullhead from the Great Lakes, in English sole from Puget Sound, and Atlantic killifish from the Elizabeth River in Virginia. But the champions,' he said, gesturing towards the water, 'were Atlantic tomcod from here.'

At the time, nearly half of one-year-old tomcod had tumours in their livers; in two-year-old fish it was greater than 90 per cent. Ike was studying carcinogenesis in rodents, as a postdoc at New York University, but the tomcod caught his attention. 'I was told that, if I wanted, I could play on the side with my fish. It led to a career studying tomcod and other species of fish in the Hudson.'

We were joined by Carrie Roble, an ecologist and vice president of the Hudson River Park Trust, and one of the Wetlab assistants, Siddhartha Hayes. Carrie had worked with Ike on some of the studies of Hudson tomcod. Raising her voice above the skirling helicopters, she told me that the project had been surveying the fish populations in the Hudson for the past thirty years. Pollution wasn't the only challenge the fish were facing. Tomcod prefer the cold, and can even produce an antifreeze protein in their blood. But warming waters had forced the population into a steep decline.

Siddhartha's arms were elaborately tattooed, with a spine of vertebrae running down his left and the right wrapped in curling tentacles. He smiled brightly. 'But we do still see them from time to time. We actually caught one this year. He's a little shy, probably hiding in amongst the mussels in one of the tanks.'

The tomcod is a small, slender fish with a tobacco speckle running along its flank. It keeps to the river bottom, where the concentration of chemicals is strongest. When Ike began studying them, he suspected that PCBs might be to blame for the high number of liver tumours. He discovered that the levels of PCBs in their livers were among the highest found anywhere in nature; yet despite their heavy intoxication, the fish kept swimming. Somehow, through exposure, the population had developed resistance. 'You can whomp them with tons of PCBs,' he said, shrugging, 'and there's almost no impact whatsoever!'

The reason for the Hudson population's extraordinary tolerance comes down to a change in the coding region of a single gene, aryl hydrocarbon receptor 2 (AHR2). This receptor plays a crucial role in the metabolism of xenobiotics, chemical substances that aren't normally present in the body. In Hudson tomcod the AHR2 is missing six base pairs of DNA that are present in other populations along the Atlantic Seaboard. PCBs love lipids, but once embedded in the tomcods' fatty tissue they can't gain purchase on crucial receptors. Like pages in a book where the glue has evaporated, without the missing base pairs and the amino acids they code for, PCBs attach very poorly to the receptors. Where PCBs bind tightly to the tissue of other fish, in Hudson tomcod sheets of toxins drift like loose leaves in a library.

It is likely that the mutated receptor was present in some tomcod before the Hudson was saturated by PCBs. When the chemical onslaught began, those fish that had the mutation were primed to survive. The change would have arisen swiftly. 'The more genes involved in an adaptation, the longer it takes for it to be fixed in a population,' Ike explained. 'If you've got just one gene, the evolution can occur much more rapidly.' Fish with the mutation survived, spreading the modified receptor through the population in as little as fifty years.

Ike told me that another population of tomcod in nearby

Newark Bay had evolved resistance to a different industrial toxin, tetrachlorodibenzo-p-dioxin, or TCDD, via the same mutated aryl hydrocarbon receptor. But for most of us, things tend to stick, and what lodges in our tissues is not just the chemicals but the histories that made them.

'We wear archives of touch,' Adam Dickinson had told me. TCDD was a by-product in the manufacture of Agent Orange, which was manufactured in New Jersey. While clouds of military herbicide were scorching the forests of Vietnam, dioxins from a Newark plant flooded the Passaic River and flowed into Newark Bay. These violent histories are part of my body, and yours, and very likely the bodies of all the people you love as well. Their traces are probably found in the leaves of every tree and the feathers of every bird that you have ever seen. Lamentable and permanent, they bind us together in what Canadian Métis scholar Michelle Murphy calls 'chemical relations'.

Chemical relations can include violent clashes between different ways of seeing the world. The Aamjiwnaang First Nation have inhabited the St Clair River in Ontario, Canada, for millennia. The Aamjiwnaang concept of 'Land' (as opposed to common or garden 'land') conceives of time as elastic, capable of extending and compressing in ways that connect different generations. 'Land stretches forwards and backward in time,' Murphy explains. 'Communities and humans are made of and through Land; they are themselves manifestations of Land.'

But with the arrival of the petrochemical industry, Land was obscured by a more prosaic notion of land as simply where value is extracted or waste is dumped. In 1858, in what is now Petrolia, Ontario, speculators struck black gold. Swamping nearby creeks and streams, the spill reached as far as the St Clair River. Today, the area is known as Canada's 'Chemical Valley' – home to 40 per cent of Canadian petrochemical processing. The St Clair is a key node in the Great Lakes network. It forms a natural passage between vast Lake Huron and Lake St Clair, from where it flows south into the Detroit River and empties into Lake Erie. In 1825, the completion of the Erie Canal had linked the lake with the Hudson River and the Atlantic, creating what was for a time one of the world's most important shipping corridors. In the decades to come, PCBs and mercury were both dumped indiscriminately into the canal's waters.

Toxins like these, Murphy says, are 'material forms of colonial and capitalist violence'. But with endocrine-disrupting chemicals this is violence that has lingered, calling back through generations like an echo. Between 1999 and 2003, the Aamjiwnaang birth ratio shifted dramatically: only a third of births were boys. The time lag may be explained by research into the effect of the chemical BPA – found in everything from canned food to plastic water bottles – on pregnant mice. Studies have shown that the main effects are not on the foetus, but on the eggs being formed *inside the foetus*. Chemicals ingested by their grandparents in the middle of one century may be determining the number of girls and boys born at the turn of the next. Multigenerational chemical harm is folded deep within the body; that which sticks to the bodies of one generation erupts in another.

Murphy suggests that, in a chemical world, we all live some form of what she calls 'alterlife', where entangled ecological, industrial and colonial histories tie us together across generations. Some are more densely ensnared than others; one hidden marker of privilege may be the degree of exposure to industrial chemicals. In North America, for example, indigenous communities are most likely to live downriver of environmental poisons. But these toxins touch everyone, regardless of privilege, embedding their histories of harm and complicity in our flesh.

While we were talking, Siddhartha had been gathering Hudson River specimens from a boat tethered to the pier. There were no elusive tomcod, but he beamed as, like a magician, he revealed a fog-coloured jellyfish perched like a hat on the shell of a marine snail, as well as a cluster of tunicates, snot-like filter feeders, clinging to the wire mesh of the crab pot in the shape of the Vitruvian Man, Leonardo da Vinci's symbol of the perfectly proportioned body. Among the mess of gloopy sediment, a gender-bending black sea bass gaped and glowered.

The tomcod survive by eluding the embrace of PCBs deep in their cells, making an untouchable life in highly polluted waters. We tend to think of ourselves as similarly sealed off from the world we pass through. This is the promise and the lie of modernity: that – whether owing to culture, technology or privilege – our way of life will make us impervious. But the hunger for contact displayed by the toxins we have seeded in soils and waterways, their inexorable desire to lodge themselves deep in living tissue, gives the lie to our fantasy of secession from nature. We can't walk away from what we carry inside us.

In which case, we need a new perspective on what it means to be so thoroughly and inescapably touched by the world, for good and for ill.

'The story of plastic begins', according to historian Jeffrey Meikle, 'with a material that often pretended to be something it was not.' The appeal of celluloid, an early plastic patented in 1869, lay in its capacity to assume the appearance of familiar surfaces. Celluloid knife handles mimicked the grain of wood; celluloid collars imitated the texture of linen. The more malleable thermoplastics that flooded people's homes in the postwar years did so disguised as a wide range of traditional materials: vinyl in place of leather and ceramic; rayon draperies and acrylic sweaters. The illusion they produced was precise enough to deceive the eye, but rarely the hand. A cellulose billiard ball might look like an ivory one, yet holding it would reveal the difference in weight; the imitation grain of a melamine table could not supplant the texture of wood. So plastic had to find a way to fool the hand, not by more subtle mimicry, but by disappearing to the touch altogether.

Unlike wood and stone, which wear their histories on their surfaces, plastic is strangely mute about its past. There is a subtlety to even the most garish plastic – a kind of discretion, because all plastics decline to speak of their origins. The illusion that even the most ordinary plastic object is somehow out of time, perpetually isolated in its moment of use, is fundamental to its promise to be disposable. History sloughs off their wipe-clean surfaces. The cut-price synthetics that inundated American society in the 1950s betrayed this promise when they quickly became discoloured and tacky, often leaving behind an oily residue. The more robust plastics that followed – silenced and stilled by their chemical additives – gave away nothing of where they came from, and so could easily be taken for granted and cast away.

Plastic blunts our sense of connection. Could we learn to feel the world again?

During my time in the US, I also travelled to Cambridge, Massachusetts. I wanted to meet several researchers who could tell me more about how natural alternatives to plastic might help us reconnect with the world around us. The first was Shannon Nangle. Shannon is the CEO and co-founder of Circe Bioscience, a company she set up with fellow Harvard graduate Marika Ziesack to develop carbon-neutral food and chemicals. Circe has devised a method of producing food through microbial fermentation, similar to that which George Monbiot explained to me. Rather than drawing on plant or animal fats, their method employs a microbe engineered to metabolise CO_2, converting greenhouse gas into fats, which can then be transformed into food. Prior to thinking about food, however, Shannon and Circe worked on using the same method to produce carbon-neutral bioplastics.

In recent years, the extraordinary cost of single-use plastics has become apparent. At present, between 4 and 8 per cent of the oil we extract is heated and cracked and converted into plastic, but by the middle of this century it could be between double and five times that much. Plastic production already yields around 200 million megatons of CO_2 emissions annually; what is more, some plastics continue to emit greenhouse gases throughout their long lives. Low-density polyethylene (PET), used to make shopping bags and six-pack rings, is the plastic most commonly discarded in the environment. When exposed to sunlight, it releases ethylene and methane, a greenhouse gas that traps twenty-eight times as much heat in the atmosphere as CO_2.

Marine microplastics are also reducing the capacity of the oceans to sequester carbon. So far, between a third and a half of our CO_2 has been absorbed by the oceans. But that capacity to soak up emissions is diminishing. Microalgae that ingest microplastic particles – often saturated in biotoxins, like tiny poison pills – are less effective at removing carbon from the atmosphere.

ONE TOUCH MAKES THE WHOLE WORLD KIN

Many plastics perform essential functions in medicine and engineering; they can even play an important role in sustainability, offering lightweight, more fuel-efficient alternatives to metal in cars and planes, as well as insulating buildings and reducing food waste. But nearly half the plastics we make are intended to be used once and discarded. Bioplastics, which have many of the same affordances as regular plastic without the need for toxic additives and are made from degradable, renewable resources, could help drive down the emissions associated with plastics and stem the flood of waste into the environment. Polylactic acid, or PLA, is derived from lactic acid and can be clear and rigid, making it an alternative to PET. Polybutylene succinate (PBS), in which the chief ingredient is lignocellulose, is flexible enough to replace polyethylene and polypropylene. Polyhydroxyalkanoates (PHAs) produced by bacteria and algae have perhaps the greatest potential, as they can be either rigid and brittle or soft and flexible.

I met Shannon, dressed all in black with heavy boots and a sharp undercut, at Circe's headquarters. We walked a few blocks away to get coffee, then returned to the office to talk.

Shannon had an entrepreneur's energy and a scientist's enthusiasm, but to my surprise what she most wanted to talk about wasn't the future of plastic, but the myths of ancient Greece. Circe stands for Circularizing Industries by Raising Carbon Efficiency. But Shannon said she was also drawn to name the company after Circe, the daughter of Helios, Greek god of the sun. Circe is represented as a maker of bitter poisons in Ovid's *Metamorphoses*. 'I would say Ovid gives Circe an unfair characterisation,' she said. 'I look more towards Homer.'

Ovid's Circe is vindictive and cunning. Jealous of Scylla, who is loved by Glaucus, Circe taints the pool Scylla bathes in with 'wonder-working poisons', which transform the lower part of her body into a writhing mass of 'barking monsters' and 'gaping mouths'. But in Homer's account, in *The Odyssey*, Circe shows another side. When Odysseus and his men first arrive on her island, she enchants the scouts Odysseus sends ahead, transforming them into pigs. Odysseus escapes the same fate with the help of Hermes, the gods' messenger, and his resistance to her magic prompts a change in Circe. 'I am amazed that you could drink my potion / and yet not be bewitched,' she declares. Circe invites him into her bed, so that 'through making love / we may begin to trust each other more'.

'So Circe shows up in the dead middle of the poem,' Shannon explained, 'which is the turning point for Odysseus. He's up shit's creek, it's his darkest hour. But Circe is one of the most empathetic of goddesses. Her first reaction when presented with something different is to say, "Let's establish trust. Let's figure out, where do we go from here?" She turns everyone back to people – they're hotter than before! – and for the next year she rests them, heals them, feeds them. So much about her house is rejuvenating. And then she tells them how to get home.'

Shannon shrugged. 'She's a magician who can transform the natural world. And what we remember from this story is that she is a temptress that turns men into pigs.'

Circe (the goddess)'s willingness to trust has informed every stage of what Circe (the company) does, including the way they work with the microbes. 'We're using the microbes as a tool, but we need to build the metabolic engineering in such a way that they can grow happily. So there's a lot of relationship-building with the microbe,' Shannon laughed. When they were working on bioplastics, they engineered microbes that metabolised carbon dioxide and hydrogen to create PHAs, biodegradable fatty-acid polymers that have similar properties to polypropylene. Other microbes can munch on the fats in PHAs when they've reached the end of their useable life, creating a circular, microbe-to-microbe economy.

But Circe stepped back from manufacturing bioplastics, not because of a limitation with the technology, but because of the sensory challenges involved. Bioplastics don't behave exactly like fossil fuel plastics; there's always a trade-off. Where fermentation can produce food that looks, tastes, smells and feels just like animal foods, bioplastic is 'a much poorer mimic', Shannon explained. Even if we don't contemplate it much, with plastic – as with food – there's such a strong, underlying sensory relationship. People found the materials unsettling. Bioplastics, it seemed, were a little harder for many people to trust. 'You don't notice it, but if it looks and feels different, suddenly you do.'

If we're to adopt bioplastics, we need to think about the kind of sense-world that will follow: one that will very likely be louder and more disturbing to our senses than the one we are used to. Bioplastics will not mould as readily into the shapes of our desires. They may look different, and feel different to the touch. Without flame retardants like polybrominated diphenyl ether off-gassing from their surfaces, they won't produce the characteristic 'new car smell' we associate with many plastics. Designed to biodegrade, they will exist in time in a way conventional plastics appear not to. At least at first, our bioplastic future could be filled with materials that seem otherworldly; or, given how plastic seems detached from its organic origins, perhaps the trouble with bioplastics is that they are too evidently of this world.

I said goodbye to Shannon, and made my way along Cambridge's Main Street towards my next meeting. The Mediated

Matter laboratory occupies a shared workspace on the corner of a quiet, tree-lined road. In 2014, under the direction of futurist designer Neri Oxman, Mediated Matter began work on a biodegradable polymer they call 'Aguahoja', meaning 'water-leaf', comprising cellulose, chitin and pectin, three of the most abundant natural polysaccharides. 'Derived from the sea and returned to the soil,' the website declares, 'we utilize decay as a design feature.' To illustrate this, Mediated Matter manufactured a series of immense Aguahoja chrysalises. The structures are deeply strange, like relics from an earlier age of the Earth, some distant period of monstrous gigantism; yet they are also surpassingly beautiful. Amber panels of dark-veined organic matter are held in tension by creamy veins of calcium carbonate, the whole shape curved and open like a pod that has released its seed. There have been three iterations of the Aguahoja sculpture so far; the latest version, *Aguahoja III*, is five metres tall and is composed of 5,740 fallen leaves, 6,500 apple skins and 3,135 shrimp shells. In photographs it glows with an eerie vitality.

I had fervently wished to see one of the Aguahoja chrysalises in person. Despite their natural origins, I imagined that standing before them would be like meeting an alien life form. But they were all held in the San Francisco Museum of Modern Art, 3,000 miles away, and my trip didn't allow for a visit to the West Coast. However, Nic Lee, one of Aguahoja's designers, promised to show me some test pieces he'd made.

Nic walked me through the open-plan lab space. 'The whole philosophy comes from looking at nature,' he said, 'where you oftentimes see media that we would consider to be very poorly suited for particular roles being used to great effect. A good example of this is chitin. Chitin is a very brittle material, yet nature turns that into dragonfly wings, the spongy tissue in mushrooms, and the hard shells on crustaceans.'

ONE TOUCH MAKES THE WHOLE WORLD KIN

'Just about any plastic product you could pick, you could probably find an organism out there in the world that's making something that's pretty comparable.'

Nic's desk was in the far corner, surrounded by shelves displaying what looked like magnified versions of the kind of things my children would collect on forest walks when they were small. He pointed to what appeared to be a pair of ribbed papier-mâché beakers. They were a composite of chitin, cellulose and coffee grounds, he said. Later they would add a coating derived from mushrooms that would make the material waterproof. They felt as light as paper, and reminded me of the paper combs of an old wasps' nest I had once found in our roof. But where the combs had crumbled to dust when I touched them, the composite felt as firm as it was weightless. The pieces were a little brittle, Nic told me, but adding the biopolymer lignin would toughen the material, which could be cast, like plastic, in all manner of different shapes.

On the shelves below were several wing-shaped panels, each as long as my arm. These, Nic said, were Aguahoja prototypes, their shapes modelled on butterfly wings. One honey-coloured wing had a network of creamy veins; another had darkened to a rich umber. I lifted the darker one in both hands. There was something odd about the way the panel caught and held the light. It burned and cooled, flashing brightly one moment, receding into shadow the next.

'They're a little bit further along in their ageing process,' he explained. The sculptures absorb water over time. An opaque white layer forming on parts of the surface was salt precipitating out of the chitosan, a derivative of chitin. Time was built into the structure, which was designed to react with its environment just as the component parts do on a forest floor or seabed.

I lifted one of the panels from the shelf. It was rubbery and pliable but evidently strong. The obviously organic nature of the material created a strange dissonance: on the one hand, I couldn't help a slight shiver of revulsion; but I was surprised too at how much I wanted to feel its leathery texture between my fingers. In a curious mixing of the senses, touching it evoked a kind of congealed citrus tang. I could imagine popping a piece of it in my mouth and chewing.

Biopolymers would not replace all plastics. Window frames and dashboards, sailboats and sunglasses, where biodegradation would be problematic, are likely to continue to be made from conventional synthetic polymers. But there are a wide range of applications where bioplastics could do as good, or an even better, job. PLA can replace PET in food packaging, eliminating harmful BPAs; biopolymers could replace plastics in agricultural mulching films and seed coatings. Chitosan has antimicrobial properties and can be used to make degradable medical dressings. As an effective flocculent, it can also be used in water purification, attracting and containing heavy metals and fine particles. Microcrystalline cellulose could replace microbeads in cosmetics; hemicellulose, or polymerised sugars, could provide films for packaging liquids. Rechargeable batteries made from crab shells (chitosan again) or seaweed, and computer chips made from mycelium skin (which can last for a century and biodegrade in a fortnight) could help reduce e-waste.

There is also the matter of scale. The world's ecosystems can produce 80 gigatons of cellulose in the time it takes us to manufacture .03 gigatons of plastic. But one study suggests that to replace just the 170 megatons of plastic packaging that the world consumes annually with organic matter would require a land area the size of France and 390 billion cubic metres of water (60 per cent of what Europe uses each year). Mediated Matter chose to design with waste rather than farmed materials. By doing the same, the bioplastics industry could become one of the world's greatest restoration movements, planting forests to harvest leaf fall and restoring marine ecosystems to collect shrimp shells. A study in *Nature* argued that a circular bioeconomy could turn the bioplastics industry into a net carbon sink, recycling what biodegrades while locking away up to 75 gigatons of biogenic carbon in nonbiodegradable materials. But such a circular economy would depend on us all radically reducing how much plastic we consume.

Perhaps this is where the compelling strangeness of Aguahoja intervenes. 'There are limitations to what we can do with cellulose right now,' Nic conceded. 'We can't just take fallen leaves off the ground and turn them into plastic bags.' But for him, the real barrier is in our imaginations: 'Whether or not we are willing to change our conception of what those products are or what they should be.'

According to the philosopher Walter Benjamin, 'all disgust is originally disgust at touching'; specifically, it arises when we are confronted by our connection to other living things. 'The horror that stirs deep in man is an obscure awareness that in him something lives so akin to the animal that it might be recognised.' We recoil from contact with another animal because it might lead to the recognition that we, too, are animals inevitably caught in a process of decay. It is this horror that plastics, with their wipe-clean surfaces, are designed to assuage. We trust plastic because it does not confront us with our own leaky, changeable, animal nature. When it arrives in our hand, it comes without attachments or history, its hydrocarbon origins neatly suppressed; even when they eventually perish, most plastics bleach into anonymity.

As Shannon said, it is a matter of where we find trust in the materials we use. An obviously organic substance like Aguahoja might not offer the reassurances of plastic, but there is a deeper trust to be found in handling such materials. As Benjamin understood, touch leads to recognition; but that can be a source of hope rather than disgust. Aguahoja practises none of plastic's deceptions. Plastic offers itself to us 'unworlded', but biopolymers would restore the world to us, affirming our place in the rich cycle of flourishing and decay. Touching a substance like Aguahoja – in the form of packaging, as cutlery, or in the stem of a toothbrush – would be a revolution in our day-to-day sense-world. The myth that we live apart from nature would dissolve in a life filled with this kind of material. Where plastic is annihilated in our imagination whenever we dispose of it, the weirdness and wonder of the living world would greet our every touch. Who then could tell where it ended and we began?

After I'd said goodbye to Ike, Carrie and Siddhartha at Pier 40, I left the river behind and headed east towards the Bowery. On an empty downtown street, a poster for an art exhibition in Chelsea caught my eye. 'What are we becoming?' it asked. 'A species between worlds.'

Our chemical world is full of substances that can intervene in us from without, diverting cells from their original path or warping them into cancerous shapes. We are each of us, plant or animal, a spectacular and horrifying crowd of synthetic compounds that can edit and alter what our bodies become. But Michelle Murphy's alterlife also contains the conditions for another kind of life and another kind of world. After all, 'life already altered', she writes, 'is also life open to alteration'. Species like the round goby and tomcod live a form of alterlife, finding

ONE TOUCH MAKES THE WHOLE WORLD KIN

ways not only to thrive in a chemical world but, in some cases, to offer a glimpse of that same world restored. Biopolymers could transform our chemical world, if we surrender our senses to what will, at least initially, seem alien and strange – and discover, as Shakespeare once wrote, that 'one touch of nature makes the whole world kin'.

Megan Watts Hughes, three-pitch Impression Figure ('Octave and 5th interval B flat') made with an Eidophone

4

THE KINSHIP OF LANGUAGES

How animal song can teach us to listen to other species

In November 1990, a cerebral haemorrhage robbed the Swedish poet Tomas Tranströmer of the power of speech and the use of his right arm. The diagnosis was severe non-fluent aphasia with dysgraphia. Although he continued to write poems, Tranströmer's primary mode of expression thereafter was music; specifically the piano, played with his left hand. When he was awarded the Nobel Prize in Literature in 2011, rather than give a speech he performed a piece of music composed for the occasion. For the final twenty-five years of his life, Sweden's greatest living poet found himself, in a phrase lifted from a poem he wrote a decade before his stroke, possessed of 'language but no words'.

Poetry occupies 'the crevice between sound and language', according to Japanese writer Yoko Tawada. It is both the chasm and the bridge that spans it. Its beginnings were in music, in chant and song; the lyric poem, which held sway in the European tradition for centuries, originated in *melos*, classical Greek poetry that was composed to be sung, accompanied by a lyre. Gradually, the music was absorbed by the speaking voice. Tranströmer's stroke teased voice and music apart, stealing the one but leaving the other. Yet his post-stroke poetry is filled with images of quietened musical instruments: a violin immured in its black case; a silent organ; a dream in which the poet sketches a run of piano keys on his kitchen table and 'play[s] on them, silently'. Animal sounds take the place of human music: stock doves calling, frogs singing, gulls screaming. It is as if, for all it took away, Tranströmer's aphasia also gifted him an affinity with those whose voices sound without speech.

Biologists recognise culture wherever a creature's behaviours meet three criteria: that they are learned; that they are distinct; and that they demonstrate some degree of stability over time. Animal song meets all three. Many animals make sounds, but for most these are genetically coded, little different from pre-uploaded recordings. For a select few, however, song is a tradition that must be taught. Both baleen whales and songbirds are vocal learners, which means they must acquire their songs from others – either vertically from parents, or horizontally from peers. In most cases, learning is limited to a brief sensitive phase when, like human children, both baby whales and juvenile songbirds will babble a form of 'subsong', a precursor to mature communication, mimicking the sounds of experienced singers until their own song crystallises. Juvenile birds will perform their subsong by listening, first to older singers and then to their own stumbling imitations. The bird listens until its body is filled with song, its whole being vibrating in synchrony – until it is ready to venture into sound and enter a community of voices.

In a very literal sense, we are muffling or warping many of these creaturely songs. Savannah sparrows in the US have evolved distinctly different songs as a result of living alongside humans. As their numbers dwindle, critically endangered birds like Australia's regent honeyeater miss the opportunity to learn vital courtship songs from their peers, in a phenomenon that has been compared to the collapse of human linguistic diversity since colonialism. Animal cultures are a kind of social glue, and species with rich and robust shared traditions will enhance the overall fitness of the group if individuals can find mates who are better at this kind of social learning. Culture is a driver of evolution, operating as a secondary inheritance system that bears hard-won advantages through time. When we weaken the song, we interrupt that flow. The song of life is haunted by an advancing aphasia.

Aphasia can take various forms. Expressive aphasia affects communication; receptive aphasia impacts comprehension. Non-fluent (or Broca's) aphasia robs the sufferer of speech, as in Tranströmer's case, but there is also a fluent aphasia, called Wernicke's (like Broca's, named for the affected region of the brain), in which language becomes a flood pouring uncontrollably from the speaker without logic or reason. Ironically, Wernicke's resembles the babbled speech of young children, as if in breaking down, language returns to its origins. The philosopher and translator Daniel Heller-Roazen notes that this nascent phase represents our capacity for language at its most potent: just before they acquire their first recognisable words, babbling infants can make a variety of sounds that would shame the most gifted linguist. When we first begin to mimic speech, every conceivable sound is on our tongue. Gradually, we learn which are useful and drop the rest, but for a short time all of speech is available to us.

The world over, animal communication is beginning to fragment under pressure. Unfortunately, for too long we have chosen not to listen. Our carelessness has had profound consequences. A recent study in *Nature*, which surveyed the decline in natural soundscapes at over 200,000 sites across Europe and North America over a twenty-five-year period, concluded that one of the most essential pathways humans have for engaging with nature – sound – is in chronic decline. If we are to truly learn from nature how to make a better world – how to build generous cities, feed ourselves sustainably and deal with pollution – then we need to learn to pay heed; more than this, though, we need to learn to take our place in the chorus, to join our voices with the rest of the living world. Subsongs are hymns of the widest possibilities. Can we find a way back from this point of breakdown, to a place where communication – language and song – might be learned again, together and anew?

If we are to learn from other species, then first we need to learn to listen to them. But when animals sing, are they speaking or making music?

Animal song cultures sit somewhere between what we define as music and what we define as language; or, perhaps, somewhere far removed from both. It's difficult to know precisely what significance many animal songs have for the singers and those that listen, or what this might have to do with song or speech as we understand them.

Nature's sounds, wrote the composer Igor Stravinsky, are no more than 'promises of music: it takes a human being to keep them'. But zoomusicologist Henkjan Honing has found that a wide range of animals possess the cultural and biological traits that make it possible to respond to music, including beat perception and the capacity for entrainment (to join in with a rhythm), pitch perception, and sensitivity to various spectral dimensions of music. Famous examples include YouTube sensations Snowball the cockatoo, who dances in time with 'Everybody' by the Backstreet Boys, and Panbanisha, a bonobo who was filmed improvising on a keyboard with Peter Gabriel. In research settings, pigeons have been trained to distinguish between works by Bach and Stravinsky, and even koi carp have been taught to know the difference between a Bach oboe concerto and the blues of John Lee Hooker.

Most researchers stress the functional aspect of animal song. To many biologists, song cultures – like any form of shared, organised animal behaviour – are just trafficking in essential information, emphasising sexual fitness (the animal that can sing the best song ought to make a more intelligent mate), territorial defensiveness, or social or parental bonding. But who is to say that pleasure isn't a part of this? The hearts of female Bengalese finches beat faster when they hear males sing more complex songs, and zebra finches get a hit of dopamine when they sing. Even when their songs have crystallised, settling into a stable form, some adult songbirds engage in 'whisper song', quietly singing to themselves when alone. Perhaps this is just rehearsal, keeping their memory sharp for the moment of performance, but we shouldn't discount the joy of singing.

Male humpback whales sing to one another in baroque sequences of growls and whistles (the females do not sing). Their range of expression is extraordinary, from a reverberant, bovine drone to a keening squeak like a finger on wet glass. Humpback whale song was made famous in 1970, with the release of *Songs of the Humpback Whale*, a series of recordings by bio-acousticians Roger and Katy Payne that became the trippy soundtrack to a generation's eco-awakening. The LP is credited with inspiring popular support for the 1972 international whaling moratorium. But aside from this – and providing the background to hip 1970s dinner parties – the Paynes' recordings broke open a problem that researchers are still reckoning with today: is humpback whale song music or language?

In 1971, Roger Payne published an article about their research in *Science*. The front cover of the issue for 13 August was covered in a strange notation, arranged in horizontal rows. It looked like a primitive or even alien writing system, but in fact it was based on a sonogram of the whales' song. By hand-tracing the blurred shapes produced by the machine, the Paynes realised that whale song exhibited certain features integral to human speech: just as we conjure sentences by combining smaller sounds into larger, meaningful ones, phonemes into morphemes, humpback whales build their songs from individual vocalisations, joining grunts and trills to make phrases. Combinations of phrases, often repeated, produce distinct themes, and a sequence of themes results in a song. Typically, a song will last between five and thirty minutes, but singing sessions can go on for hours.

Hierarchy is common to human language, human music, and to the song of animal vocal learners (songbirds construct their songs in the same way; in fact, a humpback whale's song will sound very much like a nightingale's if speeded up, and vice versa). Still, researchers tend to be wary of making definite claims that whale song is either music or language. It is evidently a form of communication – singing is especially important during the winter breeding season, when migrating whales from all points of the ocean converge on nutrient-rich seamounts – but if it is also language, then we don't know what specific meaning might be attached to any given unit of sound. To define it in strict terms as one or the other might be imposing a human understanding of culture on a behaviour with a multiplicity of functions, but that is primarily evolved to sustain life.

If neither language nor music quite fit, what about poetry? Is whale song a kind of cetacean verse, embodying echoes of the breach and the dive, the tail slap and the glide? The idea is perhaps not as fanciful as it may seem. In studying their transcription of whale song, Katy Payne noticed that certain features seemed to rhyme. A similar sound would occur at particular intervals as if to signal the end of one phrase and the beginning of another. The purpose of these punctuation points seemed to be to aid memory: breaking the song sequence into more digestible chunks made it easier for the whales to remember their epic sequences, just as the repetition of epithetic phrases such as 'swift-footed Achilles' and 'the wine-dark sea' helped oral poets carry long poems like *The Iliad* in their memories. In assembling their phrase-songs like the ancient bards, drawn from a common word-hoard, whales are the skalds of the ocean. But whale rhymes also connect their song to something much larger: the sense, which is the essence of rhyme, of connectedness; that a thing always has something to do with other things.

Ralph Waldo Emerson and Charles Baudelaire (both of whom succumbed, like Tranströmer, to non-fluent aphasia at the end of their lives) spoke of rhyme as the fundamental principle in nature. Baudelaire described nature's correspondences – the teeming array of mutualisms and synchronisations that sustain life – as 'these grand rhymes'; for Emerson, it is by rhyme that 'we participate [in] the invention of nature'. The humpback whale's song may be a trick that allows the animal to recall its long compositions, but it also carries an echo of the great, wide song of creation.

Subsequent studies of humpback whales have revealed that their song not only possesses a grammar and makes use of rhyme, it also changes over time. The phenomenon was first observed in the 1970s, but recent research has revealed the extraordinary extent to which evolution is a fundamental part of whale-song cultures. New elements come in continually, either because an error spreads after one whale mishears a phrase, or through improvisation. Gatherings of different pods provide opportunities for innovations to spread, leading to a continual, progressive evolution of the song over time. Something similar has been observed in songbirds: a thirty-five-year study of Savannah sparrows revealed the gradual introduction of a short repeated click in the place of interstitial notes (soft notes sung in the intervals between loud introductory notes). Click-singing males produced more fledglings, and subsequent generations added novel clicks of their own.

Of course, we shouldn't be surprised that culture doesn't stand still. In the case of some humpback populations, however, innovation is an entire way of life. Geography matters: hemmed in by the broken arch of the Kamchatka and Alaskan peninsulas, northern populations have relatively few opportunities to mingle and swap influences, but pods in the southern Pacific have developed an approach to song that is as radically expansive as their ocean home. These whales will replace their entire song catalogue in a matter of years in a process scientists call 'song revolutions'. The new song is usually less complex than the one it replaces, but complexity is gradually restored over time. Crucially, the old song is forgotten entirely.

Where we prize our heritage, attaching special value to the achievements of the past, for these whales, the past is nothing; the song being sung is all. Imagine if Nina Simone or the Beatles had deleted their back catalogue with each new record. It is unthinkable, but according to marine researcher Ellen Garland, once a whale song falls out of circulation, it is gone forever; there is no recorded instance of a dropped song being resurrected. (The speed with which revolutions take place suggests there isn't such a thing as a 'word' in whalish, which would require a degree of semantic fixity over time.) Garland

THE KINSHIP OF LANGUAGES

has tracked the astonishing spread of song revolutions in the Pacific. Between 2003 and 2005, the song of an eastern Australian population of whales, which they had acquired from a western Australian population, spread to French Polynesia, a distance of over 6,000 miles. Garland and her colleagues traced the further spread of the song and its variations as far as the coast of Ecuador – literally, a world-spanning song, continually made new.

In *Echolalias*, a study of linguistic forgetfulness, Daniel Heller-Roazen recounts a story in the life of the classical Arabic poet Abū Nuwās. As a young man, Abū Nuwās approached an older poet, Khalaf al-Ahmar, for permission to write poetry. He was first instructed to memorise a thousand lines of ancient verse; when this was done, and Abū Nuwās recited the entire catalogue to his mentor, Khalaf replied that he would withhold permission, 'unless you forget all one thousand lines as completely as if you had never learned them'. Like Abū Nuwās, humpback whales make their song by first forgetting all they once knew.

However, the oceans are getting noisier and drowning the whales' song. Since the 1960s, low-frequency background noise has risen by around three decibels every decade. Shipping is the main cause, with the greatest noise coming from propeller cavitation (the sound of clouds of bubbles erupting behind the propeller). Engine noise and the vibrations of the hull, and a phenomenon called propeller singing (the vibration of the blades), add to the cacophony.

More recently, mining is driving the racket to an even greater pitch. The Clarion-Clipperton Zone is an 1.7 million-square-mile area of seabed in the eastern North Pacific, lying at an average depth of 5,500 metres. It is an essential habitat for both baleen and toothed whales; it's also a key theatre in the race to extract polymetallic nodules – agglomerated lumps of valuable minerals like copper, nickel, cobalt, iron, manganese and rare earth elements, which litter the abyssal plains. Commercial-scale mining of these resources is hugely destructive: although they lie scattered on the seabed like lost golf balls, the depths involved mean the only way to recover them is by dredging, which scrapes the substrate clean of all life and nutrients as well as the minerals. It is also extraordinarily noisy. Every species has a designated place in what sound ecologist Bernie Krause calls 'the animal orchestra' – its own slice of bandwidth, in which its calls can be heard without the masking effect of competing species. But mining operations and humpback whales share a portion of low-frequency bandwidth (dredging causes noise between 0.02 and 1 kilohertz, and commercial drill ships between 0.02 and 10 kilohertz; the whales' song sounds in the bottom end of this range, between 0.02 and 4 kilohertz). The overlap means that mining noise cancels out or 'masks' the whale song, degrading their perception of important signals and disrupting their ability to coordinate behaviours like foraging and breeding.

It is ironic that mining for minerals necessary for communication devices comes at such a heavy cost to animal communication. The immediate risks include ship strikes (the noise field from shipping is not isotropic – it doesn't issue in all directions equally – so while a whale at the rear may be deafened, another lying in the path of an oncoming ship may be oblivious to the danger) and strandings (military sonar has been linked to an increase in the stranding of Cuvier's whales). Sometimes it seems human noise can be almost mesmerising. Foraging humpback whales have been observed behaving sluggishly in noisy waters, with a slower descent and fewer side-roll feedings, and narwhals can slow down, become motionless and sink. At other times, our racket seems to horrify them: sonar can produce a startle response, causing diving whales to surface too quickly, increasing the risk of a fatal gas embolism.

But there may be more long-term impacts too. Songs are changing to cope with a more crowded soundscape: humpback whales will raise the volume of their calls roughly in line with the rise in background noise – 0.8 decibels for every 1 decibel increase – in a phenomenon known as the Lombard effect (the same happens when we instinctively raise our voices as background noise increases). Grey, humpback, right and fin whales all shift to lower-frequency calls when confronted with human noise; sperm whales alter their clicks and dolphins modify their whistles. Sometimes, the animals fall silent: at distances of up to 1,200 metres from ships, humpbacks will stop singing altogether.

A long-term study of baleen whale earplugs – stalagmite-like accumulations of lipids and keratins which, like ice cores drilled from glaciers, archive a lifetime of chemical exposure – has shown that whales carry stress in their ears. Cortisol levels for the peak period of whaling in the last century were 50 per cent above the baseline; the level dropped considerably in the 1970s, following the whaling moratorium, but ever since it has been steadily climbing in line with the rise of marine noise and is now approaching levels last seen at the peak of the slaughter. But the whales also, now, carry stress in their song. If the underwater soundscape continues to degrade, it is likely that whale songs – which are both an evolutionary event in and of themselves, and an ancillary inheritance system for the species – will also degrade.

Whales aren't the only creatures whose song is suffering. Other marine species are also affected: pile-driving, for instance in the construction of oil platforms, disorients shoals of fish. The sound waves from seismic surveys wipe out huge swathes of zooplankton. On land, city birds are singing at higher frequencies to cope with the urban cacophony, while birds nesting near Schiphol and Manchester airports have been shown to sing fewer high-frequency notes, probably due to hearing loss that comes from being exposed to the 93-decibel roar of departing airplanes every 180 seconds. Traffic-noise stress has been shown to negatively impact embryo mortality and nestling growth rate in zebra finches. It depresses young birds' cognitive performance, so they are less able to learn their songs. Deep within their genomes, noise-stressed zebra finches have shorter telomeres, the region of repetitive DNA sequences at the end of each chromosome – in effect, where the genome 'rhymes' – that should act like end stops, protecting the chromosome from fraying or tangling.

If we are to ask whether whale song is music or language, then we must also ask how we should diagnose the condition inflicted by all this commotion. If it is music, then we might identify a case of dysharmonia. In *Musicophilia*, the neuroscientist Oliver Sacks recounts the story of Rachael Y, a composer who, following a car accident, lost the ability to hear harmony between different instruments, but could still perceive the intensity with which they were played. 'I absorb everything equally,' she told him, 'to a degree that becomes at times a real torture.' For Rachael, Sacks writes, music would 'burst apart, becoming a chaos of different voices'.

Or, if whale song is language, we might look to the sad story of the aphasic poet Baudelaire. A series of strokes in 1866 deprived him of all but a single word: 'crénom', a contraction of 'sacré nom de Dieu', which he would utter compulsively. In the end, the poet who saw all creation as a 'grand rhyme' could do nothing but curse.

*

THE KINSHIP OF LANGUAGES

In 2015, a forty-two-year-old gorilla called Koko delivered a plaintive message to humanity ahead of the COP21 UN Climate Change Conference in Paris. Koko, who had spent her whole life in captivity, had been taught to use American Sign Language. Her thirty-seven-word plea was stark. 'I am Gorilla,' she signed, in a film produced by French conservation charity Noé. 'I am flowers, animals. I am Nature. Man Koko love. Earth Koko love. But man stupid. Stupid! Koko sorry. Koko cry. Time hurry! Fix Earth! Help Earth! Protect Earth. Nature see you. Thank you.'

Koko was a remarkable primate. Not only did she know more than a thousand words of ASL, she also understood twice as many words of spoken English. She loved kittens, and playing on the recorder. Yet like many prophets, her message was not heeded: the Paris Agreement did not arrest runaway climate breakdown. But we might also wonder how much Koko herself understood it. When asked where animals go when they die, Koko reportedly once replied, 'a comfortable hole'. It seems doubtful that she would have grasped the notion of a planet-wide, intergenerational threat to life on Earth.

In truth Koko's plea was a highly edited fiction, spliced together from scripted phrases, often to sentimental effect. ('Man stupid,' she signs, before the film cuts to a pensive Koko, presumably contemplating humanity's bone-headedness; another cut and she exclaims, 'Stupid!') If anything, it speaks most articulately of our desire to communicate with other species. Koko is one of several primates who have been taught to sign, and each one is rightly considered a marvel. But can the gap between animal and human ways of seeing the world ever be bridged?

There's a poem by Robert Frost that imagines the beginning of birdsong. Birds in paradise, he writes, learned to sing by mimicking Eve's songs of praise, emulating 'her tone of meaning but without the words'. Ever after, the voice of Eve 'was in their song'; 'never again would birds' song be the same'. Perhaps, however, the influence was in the other direction. Charles Darwin speculated that the origin of human language lay in a 'musical protolanguage' that mimicked birdsong. All humans possess 'an instinctive tendency to acquire an art', he wrote in *The Descent of Man* in 1871; language emerged from our instinct to copy and learn, and our first teachers, Darwin speculated, were the birds, whose songs our ancestors copied, then combined with their own vocalisations, later adding meaning and structure.

It's possible that some versions of those earliest musical languages survive even today. Whistle languages modify phonetic systems into a series of simple, modulated whistles. More than seventy whistle languages are still spoken worldwide, translating both tonal and non-tonal speech into something much closer to birdsong. (In tonal whistle languages, whistles follow the changes in pitch with each syllable; with non-tonal languages, whistles mimic changes in resonance between vowel sounds.) On La Gomera, in the Canary Islands, Silbo Gomero is a whistled form of Castilian Spanish used primarily by shepherds to call between mountainous valleys. It has been taught in schools since 1999; there's even a website, Yo Silbo, which can teach you to communicate by whistling. If you do try to learn, however, know that birds have beaten you to it: blackbirds on La Gomera incorporate Silbo whistles into their song.

Recently, Darwin's speculations have gained ground with linguists studying the evolution of language. According to James Thomas and Simon Kirby, self-domestication – the process of selective taming that some researchers suggest was key to the development of modern humans – may also have played a crucial role in the evolution of language. Two hundred and fifty years ago, Japanese breeders created a domesticated strain of the white-rumped munia, known as the Bengalese finch. The domesticated birds were bred for their snowy plumage rather than their song, yet their song has changed remarkably. Both wild and domestic species are 'closed learners', meaning they have a limited developmental window in which to acquire their song from other birds. Where the wild munia faithfully repeats the song as instructed, the Bengalese finch shows a markedly lower copying fidelity. Its song admits all kinds of errors, mishearings and improvisations, leading to a melody that is syntactically richer and more complex. Coupling this with the increased sensitivity to communication cues demonstrated by Belyaev's foxes as they became domesticated, Thomas and Kirby suggest that domestication may be linked to 'language-readiness' – Darwin's 'instinct to learn' – in the human brain.

All of this suggests that, far from marking out humans as exceptional, the beginnings of language were in a form of cross-species exchange. The question remains whether humans and other animals can ever truly learn to talk to one another again.

In recent years, a number of research projects have applied machine learning to the problem of interspecies translation. An AI program called DeepSqueak can categorise the ultrasonic vocalisations of rats. In 2013, nine years before the release of ChatGPT, scientists working on a project called CHAT (Cetacean Hearing and Telemetry) may even have added a word to the vocabulary of a pod of dolphins. Using an AI algorithm, they identified a new click in recordings of the dolphins communicating with each other, which they recognised as a sound they had previously trained the pod to associate with sargassum seaweed. The dolphins appeared to have assimilated it into their own click-hoard – the first and, so far, only time humans have introduced a new word across the species barrier.

In lieu of a full cross-species vocabulary, some have resorted to more imaginative means. In 1992, the Australian poet Les Murray published a sequence of poems called *Translations from the Natural World*. From bats to lyrebirds, cattle to single-celled organisms, Murray's poems seek a language that expresses each being's distinctive way of seeing the world. In 'Spermaceti', the poem's sperm-whale speaker conjures a realm whose parameters are determined by shifts in buoyancy and sound, fluctuating densities and deep breaths, in which 'every long shaped cry / re-establishes the world'.

Murray's lyric experiments tell us more about poetry's capacity to crack open human language than anything we might hope to learn about what animals have to say. But a project called CETI (Cetacean Translation Initiative) aims to make trans-species communication a reality by translating sounds made by a population of around thirty sperm-whale families near Dominica. Sperm whales communicate through a variety of clicks, some of which are so brief – as short as one-thousandth of a second – and closely spaced, they sound like a door creaking on its hinges. Most intriguingly, they combine patterns of clicks in arrangements called codas, which seem to have a largely social purpose. The most intense codas are exchanged when groups of whales socialise at the surface. There is evidence of turn-taking, as if codas are either duets or conversations. Sperm whales refer to one another with specific clicks, and can even be divided into vocal 'clans' defined by dialect. In this case, culture creates stronger bonds than either genes or geography: whales with very similar genes can have different dialects, and clans can be distributed over thousands of miles. The loudest clicks can reach 230 decibels, way beyond the level that would

THE KINSHIP OF LANGUAGES

deafen humans. In *Moby-Dick*, Herman Melville suggested that the sperm whale speaks in strange rumbles because he talks 'through his nose'. More precisely, they achieve such a tremendous volume of noise by channelling sound through the immense spermaceti organs mounted on their heads. These cavities contain almost 2,000 litres of spermaceti oil, acting like an amplifier and allowing the animal, as Murray puts it, to 'sing beyond the curve of distance'. (Although, strictly speaking, sperm whales do not sing.)

Beginning in 2020, CETI has established a range of listening stations where the Dominican pod congregates, using hydrophonic arrays, robotic fish, and hydrophones dropped by drones to gather the sound of the whales' codas. They estimate they will need to gather 4 billion clicks to form a viable dataset for AI to analyse (ChatGPT-4 was trained on more than a trillion parameters). Their aim is to translate sperm whale by 2026. So far, they have identified around twenty-five distinct codas, each containing hundreds of thousands of individual clicks. Some may act like morphemes, assembled in different arrangements to form what might equate to words. The researchers have even isolated a particular click that may be a form of punctuation.

Still, there is much we don't understand about how sperm whales communicate with one another. Perhaps changes in frequency carry particular meanings, like the pitch changes in tonal languages. A complete sperm-whale grammar eludes us.

The massive computing power of intelligent machines may crack that code, eventually. But there are greater difficulties beyond language rules. Even with creatures that, like us, use sound to communicate (as opposed to a more remote form of communication such as chemical cues), we can't presume to understand what role sound plays in their experience. Where we pay attention to melody when we hear a piece of music, birds ignore melody entirely in favour of timbre, attending to minute variations in tone and colour or the energy with which a note sounds. We can recognise a melody regardless of the instrument on which it is played, but a songbird will assume a melody played twice with changes in timbre, or at a higher or lower pitch, represents two entirely different songs. ('One could say that songbirds listen to melodies the way humans listen to speech,' Henkjan Honing observes, given that inflections in tone matter much more to us when we talk to one another than when we sing.) In honeybee colonies sound is only one element in a richly 'noisy' environment, in which chemical, mechanical and thermal cues also play a part. For both dolphins and sperm whales, sound is a vastly different sense than it is for us. Both species can translate the sound waves that reflect off surfaces and come back to them into images of their environment. Echolocation or visual hearing blurs several senses together, and echolocating animals separated by distance can even 'see' what their fellows see by tuning into the reflected sound waves.

To truly understand what any creature communicates, we would need to be immersed in their world as they experience it. In 1909, German biologist Jakob von Uexküll proposed that every living thing inhabits a perceptual world that is distinct to its species, composed exclusively by the information the animal can sense. To a dog, reality is a concoction of different smells; to a spider, the world vibrates. To dolphins and sperm whales, we might imagine it shimmers with the ghostly impressions of their echolocation. 'Earth teems with sights and textures, sounds and vibrations, smells and tastes, electric and magnetic fields,' writes the science journalist Ed Yong. 'But every animal can only tap into a small fraction of reality's fullness.' These radically different ways of sensing the world naturally lead to incompatible modes of communication. What prevents us from talking with other species is not just the absence of a common language, but the fact that our biology imprisons us on wholly different worlds, as alien as if one participant in the conversation had arrived from another planet. It's no coincidence that CETI was named after NASA's SETI (Search for Extraterrestrial Intelligence) programme.

Von Uexküll's term for these distinct and immutable perceptual worlds was 'Umwelten'. Figuratively, an Umwelt is often spoken of as a bubble world, an image that originates in a thought experiment he proposed in 1934. On a stroll through a meadow on a sunny day, we might glimpse the busy lives of its inhabitants – flowers, butterflies and other insects – but to do so, von Uexküll suggests, 'we must first blow, in fancy, a soap bubble around each creature to represent its own world, filled with the perceptions which it alone knows'. We can see the butterfly immersed in its world of sensation but cannot reach it. It's an evocative image, if an oddly ephemeral one. Bubble worlds in their impregnability are more like iron strongboxes. And yet perhaps the point is not how robust the boundary between worlds is, but how well that world can withstand intrusion; to enter the world of another species, we must prick the bubble.

'If a lion could talk,' writes the historian Stephen Budiansky, revising Wittgenstein's famous aphorism, 'we probably could understand him. He just would not be a lion anymore; or rather, his mind would no longer be a lion's mind.' This is the danger and the opportunity of interspecies communication. If we truly were to succeed in exchanging understandings of the world across the chasm of our respective Umwelten, what of us might be impressed upon the animal we speak with, and what of them might, in turn, be impressed on us? Other species already have to contend with a host of human pressures intruding on their minds and bodies. Marine noise threatens to warp the evolution of humpback whale song cultures, adding an unwelcome strain of human dissonance. In breaking into their bubble worlds, perhaps we would risk compromising further what it means to be that animal. But there is also the possibility that animals teaching us their speech might change us. If we could talk to and understand a lion, not only would his mind no longer be that of a lion; our minds too would surely be altered, carried across the species gap to the edge of something that is no longer only human.

In Ted Chiang's short story 'Story of Your Life', a visit from intelligent aliens is the occasion for just such a transformation. Chiang's alien 'heptapods' look like 'barrel[s] suspended at the intersection of seven limbs' and seem relatively human-sized. In *Arrival*, Denis Villeneuve's film based on Chiang's story, they have a cetacean massiveness, towering over their human interlocutors and speaking in a mixture of booms and creaks. In both film and story, the protagonist, Louise – a linguist responsible for establishing a way to communicate with the aliens – realises that their spoken language, dubbed 'Heptapod A', is unpronounceable; but their script, 'Heptapod B', offers a window onto their reality.

Heptapod B, she discovers, is a 'semasiographic writing system', in that it bears no relation to sound. In *Arrival*, it is formed by a blue-black substance that coalesces in a messy circle when squirted from one of the alien's tentacles, with branching stalks and kernels like a Rorschach blot; in Chiang's story, the semagrams resemble praying mantises or spiderwebs. Most startlingly, Louise realises that the heptapods write by layering words together, producing long, complex sentences instantly, almost as if they know in advance what they are going to say – an insight that is the catalyst for a profound change in how she perceives reality. Where humans experience time sequentially, expressed in the linear organisation of our sentences, for heptapods all time, like their script, is simultaneous – or, at least, they make no distinction between remembering the past and remembering the future. As semagrams blossom in her mind's eye like mandalas, they lift Louise out of sequential thinking and into a memory-scape where past and future are equally available.

While Chiang's story doesn't address directly why language should have this effect, *Arrival* offers the Sapir-Whorf hypothesis as justification. In the early decades of the twentieth century, linguists Edward Sapir and Benjamin Whorf argued that the experience of reality is encoded in language: how one speaks influences – even defines – how one thinks. 'We dissect nature along the lines laid down by our native languages,' Whorf claimed. Their argument was based on a study of indigenous American languages – especially that of the Hopi, who, it was alleged, lacked verb tenses for time (giving rise to the urban myth that the Hopi do not perceive time). According to this theory, learning Heptapod B – or any non-human speech – could have a profound impact on how the speaker then sees reality.

The Sapir-Whorf hypothesis was discredited by later generations of linguists, but more recently scholars have argued for a less strident version of linguistic relativity. Where English speakers conceive of time horizontally, Mandarin speakers imagine time as vertical. Among the people of Pormpuraaw, an indigenous community in Cape York, Australia, time is not arranged according to the body of the speaker (as in both English and Mandarin), but according to cardinal directions: specifically, time moves from east to west. In Kuuk Thaayorre, the language of Pormpuraaw, 'time flows from left to right when one is facing south,' writes linguist Lera Boroditsky, 'from right to left when one is facing north, toward the body when one is facing east, and away from the body when one is facing west.' In other words, Pormpuraaw time is indivisibly linked to the relationship between body and land.

Some indigenous Australian languages (Murrinhpatha, which is spoken along the northwest coast, and Pitjantjatjara, which is spoken in the central region) have free word order, which means subject, object and verb can occur at any point in the sentence. Moreover, Murrinhpatha is polysynthetic, which means each sentence comprises a single word assembled from many affixes (and every sentence is a word newly minted for the occasion). Whereas speakers of a language with a fixed order of subject, verb and object, such as English, will fix their eyes solely on the subject of their sentence before they speak, studies have shown that the eyes of Murrinhpatha and Pitjantjatjara speakers move rapidly – in the order of milliseconds – between every component before settling on a word order for their sentence. Living in a language with free word order literally affects the way speakers look at the world, allowing them to frame the whole picture and the relations contained within it before venturing to speak.

The technical term for this is 'weak linguistic determinism' (in contrast to the strong version advocated by Sapir and Whorf). It suggests that language indeed plays a limited role in shaping our reality, if not a wholly determining one. In which case, we might legitimately speculate what effect learning to speak whale – a form of communication so alien it is more akin to Heptapod B than English, Mandarin or Kuuk Thaayorre – would have on our human Umwelt.

Sperm whale codas and humpback songs are born out of an experience of time and space that is radically different to our own. Learning to speak whale would mean adopting an ocean-going animal's sensibility. A humpback whale can communicate directly with others of its species up to a distance of around two and a half miles in every direction. Within that range, any song can be heard by one or many intended receivers, as well as any number of unintended receivers. The space into which whales speak is more expansive than ours in every sense, both larger and more inclusive, yet the clamour of our seagoing vessels can reduce the communication space between whales to just half a mile, limiting the audience to a single receiver.

Humpback whales can sing for up to twenty-three hours in one session. Their songs travel the world's widest ocean, evolving as they go. Imagine what joy must infuse a language that inspires its users to such feats – the swell of oceanic feeling on which its sounds are borne. Speaking whalish with this degree of intimacy would expand our sense of space and time into a planetary song. If we were furnished with whalespeak, I doubt we would be content to pollute the ocean soundscape or privatise the seabed as we do.

If Chiang's story offers us the dream of interspecies communication, then Laura Jean McKay's novel *The Animals in That Country* offers the countervailing nightmare. In the novel, a virus called 'zooflu' affords those it infects with the sudden ability to understand all animal speech and to be understood in turn. For the most part, sufferers can hear mammals, but in the most severe cases, birds and even insects become intelligible. Animals communicate with their whole bodies, different elements often working in concert, leaving 'trails of glowing messages . . . In stench, in calls, in piss, in tracks, in blood, in shit, in sex, in bodies.' Each shiver, twitch and puff of scent speaks.

McKay's novel suggests that, when it comes to interspecies communication, we should be careful what we wish for. Zooflu would be catastrophic for a society that depends on closing our eyes and ears to what animals say to us. We might think that, if we really could communicate with other species, then we couldn't help but change *everything* about how we treat other beings. All those animal claims for space to make a home in places where we want to farm or build or mine, the clamour to live coming from the slaughterhouse, would surely be impossible to ignore. But in the novel, hearing from other species only makes people more fearful and more hostile, stuffing their ears with paper to silence the noise and establishing 'people towns' where all non-human life is evicted. Cruelty and indifference towards animal suffering continues unabated. *Arrival* figures the alien language as a gift; McKay's novel suggests it is one most of us would shun. For the most part, people view their new ability as a curse: in desperation, some resort to trepanning as a DIY cure; others take their own lives. Eventually, some zooflu sufferers lose human speech altogether, becoming lost in a sense-world of animal yammering.

Those that don't succumb to fear find the language of animals to be fatally enchanting, as in one of the novel's most powerful scenes, when hundreds drown while listening to a pod of southern right whales, drawn under the waves like sleepwalkers. 'Welcome home,' the whales sing, in refrains that sound like a 'mother's voice on the other side of the womb':

> Here is where it
> came from and
> here
> is where it sleeps.

When asked 'what would you like to ask another species?' the scientists working on interspecies communication often reply with a version of the same question: *what do you think of us?* But for the most part, the animals in McKay's novel seem not to think much at all, proving to be as immersed in their worlds as we are in ours. Their preoccupations are largely fear, food, territory and sex; in other words, with life and how to live it. A wallaroo's spray of urine shouts 'Fuck me I'm / a king', while another, unseen, animal's piss yells 'Mine, / mine, mine!'; a blowfly's whine discloses a constant, two-beat mantra of 'fuck' and 'suck'. A maltreated horse eyes watchfully as 'the angry / world comes / closer'. With its contorted syntax, weird line breaks and gnomic imagery, the whole-body speech of animals yields a strange, almost aphasic poetry. But there is no great wisdom to be gleaned, no message from the Earth. 'I thought they'd have more to say,' one character complains.

But really, do we need to learn any more than we know already? Other species already bear eloquent witness to the distress we're causing them. A recent study has shown that animals in the wild will flee the sound of a recorded human voice with greater alacrity than even the sound of an apex predator like a lion. Every time an organism expands its range due to climate change, it brings a warning of calamity as acute as Koko's. What more could they have to say than *look at what you're doing?*

The lion speaks, as do all his many kin, and where it counts we are perfectly able to understand what they say; the problem is that we choose not to.

For now, interspecies translation remains out of reach. But no act of translation, however deft or attentive to nuance, can carry a language entire across the gap that separates it from another. Whole worlds of association and common use lie insuperably between human tongues. Why should we imagine that it would be any different when the languages are also divided by species? But there may be other ways to approach the problem of translation.

'Translation does not find itself in the centre of the language forest but on the outside facing the wooded ridge,' wrote Walter Benjamin in an essay on translating the work of Charles Baudelaire; 'it calls into it without entering, aiming at that single spot where the echo is able to give, in its own language, the reverberation of the work in the alien one.' The translator stands on language's edge, calling into its muffling depths and hoping that what returns carries an 'echo of the original'. Much is lost as the sound travels into and out of the forest; but then, in their relation to the original, translations are often more like metaphors than exact copies. Or they are like rhymes between different modes of speech. Translation's purpose, Benjamin states, is not to find exact renditions of meaning as much as to express 'the kinship of languages'.

On a walk along the jagged coastal paths of Devon, the poet Holly Corfield Carr caught a glimpse of the kinship Benjamin was writing about. 'I went to East Soar,' she told me, 'and walked along the cliffs, when I was interrupted by the rudest, loudest jay.' The bird was hidden from sight, deep in a hedge, so that only its voice announced its presence. At first, Holly said, she couldn't work out where the sound was coming from, or whether it was an alarm call or the bird was singing for some other reason. It was as if the sound was issuing directly from the stony ground.

Holly became fascinated by sonograms of birdsong for the resemblance of their dark, jagged lines to the striated, stratified landscape in which she had first heard the jay. *Subsong*, the collection of poems and field notes she wrote in response, explores this strange kinship. East Soar's cliffs include some of

THE KINSHIP OF LANGUAGES

the oldest rocks in the UK, mile after mile of glinting mica schist; and just as 'schist' means 'to split', the poems were, she told me, 'an experiment in whether you could split the voice' between the birds and the land they inhabit:

> In the rock, echoes take place
> upwardly from that one dark line
> so exactly like the shock of a jay

A split voice is the natural condition of all songbirds. Their throats have a Y-shaped confluence called a syrinx, which allows them to sing two different pitches simultaneously. They are, literally, double-voiced, and can comfortably accompany themselves in a way humans can only do with the help of an echo. As Benjamin imagined translation as a voice speaking into the centre of a forest, seeking the reflective surface that would send it back again, Holly was interested in how a voice, once sent out into the world, comes back changed by what it encounters. 'A word leaving you can't come back the same,' she told me. Somehow, it mingles with what it meets. Shout into a cave or a rushing river, and you might imagine you're playing a game with yourself, but the voice that returns 'is not yours anymore, it's been remoulded by whatever it's come into contact with' – shaped by the acoustics of rock, wood, water, or even other bodies. This doubling, hearing your own voice made strange, she said, breeds a different type of companionship with the world.

'Rhyming, for me, is generative,' Holly told me, but behind the rhymes in her poems there is a deeper harmony, the intuited rhyming – one that is seen as much as heard – of elements that seem deeply unlike: the enduring presence of the land and the ephemerality of birdsong. In the sonograms of birdsong,

'I found sound and vision tumbled together,' she writes in *Subsong*'s preface.

Perhaps, if we are to truly access the kinship of languages, we may need to look beyond sound alone; not just split the voice (and the ear), but also splice it with other ways of experiencing the world. Before he succumbed to aphasia, Baudelaire expressed the world's intricate rhyming through a mingling of the senses:

> Like echoes long that from afar rebound,
> merged till one deep low shadowy note is born,
> vast as the night or as the fires of morn,
> sound calls to fragrance, colour calls to sound.

Synaesthesia (from the Greek *syn*, 'together', and *esthesia*, 'sensibility') is a rhyming of the senses. The phenomenon affects about 4 per cent of people, for whom senses can overlap in very peculiar ways, from hearing scent or tasting sound, to feeling pain as a series of coloured tints or the prickle of sound on their skin. For each person the sensation is different, and so integral to the way they see the world that it can be difficult to put into words, which only makes it seem more spectacularly strange to those who don't share it – imagine tasting mint whenever you touched metal, or that stubbing your toe produced a vivid burst of yellow.

In the most common form of synaesthesia each letter of the alphabet has its own colour (the technical name for this is 'colour-grapheme synaesthesia'). This is one of two kinds of synaesthesia I experience, and it is also how Vladimir Nabokov saw the world. 'I present a fine case of coloured hearing,' he wrote in his memoir, *Speak, Memory*, although he conceded that 'perhaps "hearing" is not quite accurate, since the colour sensation seems to be produced by the very act of my orally forming a given letter while I imagine its outline'. In my own case, 'intuition' seems closer to the truth than 'hearing': I can't remember a time when A wasn't pale green or B an orangey-brown, hues that accompany the letter whenever it comes to mind. For some the tones have an acrylic brightness; for me they are pale or muted, like watercolours; but for everyone with coloured hearing, the colour associations are unique. Where Nabokov's C mixes azure and mother-of-pearl, mine is the green-tinged yellow of an underripe banana; his G is black as vulcanised rubber, while mine is light and warm, like a glowing hearth; his creamy D opposes mine's antique green, and so on (although both our Os are bone-white with a sharp, dark outline). I find it impossible to imagine things could be otherwise; colour and letter are indivisible. Letters in the 'wrong' colour leave me feeling ever so slightly queasy.

Relatively few people experience a mingling of perception in this way, but to most other species it is the norm. Ed Yong cites the platypus's ability to mingle touch and electrical sensations in its bill, the way an ant's antennae are sensitive to smell and to touch, and the merging of sound and vision in echolocation, as just a few of the countless ways animals merge their senses to build a picture of the world. Which suggests that, rather than direct translation, cultivating a mixing of the senses may be a more effective way to enter into kinship between languages.

While it's relatively easy to explain the colours I associate with letters, the other form of synaesthesia I experience is more difficult to describe. It has to do with seeing sound, especially music, although any sound can manifest in this way if I give it my attention. As with Nabokov's 'coloured hearing', however, 'seeing' is a little misleading. Sounds appear and fade in my mind's eye as ghostly, slightly fuzzed shapes against a dark screen. When I listen to Duke Ellington and John Coltrane's version of 'In a Sentimental Mood', for instance, Ellington's minor chords ripple across the foreground of the screen, the left-hand rhythm a step beneath what the right is doing, while Elvin Jones's brushed drums flash dimly like scattered sand somewhere above and behind. The first notes of Coltrane's saxophone glissando rise like smoke but then seem to solidify – or, they assume a consistency that is somehow both light and solid at once, and *rich*, if a shape can be said to be rich – swelling and vibrating above the rest.

Really, though, it's nothing like that. It's just the best I can do with words to describe the sensation I have when I listen to music. For a long time, I assumed that what passed across the screen in my mind would never be visible in the real world. But in my conversation with Holly, she mentioned the work of Megan Watts Hughes, an opera singer and scientist born in Wales in 1842, who became fascinated by the possibility of making sound visible. 'She invented something called the Eidophone,' Holly explained, 'like a song phone. It was basically a big pipe, a bit like a smoker's pipe or a tiny bassoon.' The device projected the sound of Hughes's voice onto a powder- or seed-dusted rubber membrane; speaking or singing into the Eidophone would cause the particles to leap up and settle into a variety of often floral shapes she called 'voice figures'. 'She became quite taken with the idea that God had encoded the human voice with elemental forms of nature,' Holly said. 'She found plants, trees, all sorts of things were in her voice.'

Experimenting further, Hughes replaced the rubber membrane with a glass plate coated in blue or yellow glycerine, and developed a handheld version that could move across the plate to capture changes in pitch or even different sounds together.

THE KINSHIP OF LANGUAGES

The images she sang into being through her Eidophone, captured against a cerulean or honey background of broad brushstrokes, look more like the sounds I see when I hear music than anything I've ever come across. An individual pitch sung into the Eidophone swells like a humpback's mooning call when it blooms in my mind's eye/ear. Its edges lighten and become thinly corrugated as the sound moves outwards from the dark centre, just as my spectral sound-shapes fray at the edges. The more complex 'impression figures', as Hughes called them, are riots of plant-like forms, branching and reticulate, and – to me – exactly as if someone had reached into my mind, plucked out the confusion of sound and image, and fixed it on glass.

There is a synaesthesia-like dimension to the science of visualising animal sounds. But sonogram transcriptions of birdsong or whale song, which isolate distinct sounds and give each its own shape just like musical notation or the letters in an alphabet, only emphasise their difference from our languages. They look more like Heptapod B than any human script, as if forcing the voices of other beings into human shapes just serves to push them further away. For sure, there's a sense of kinship to be found in the realisation that animals might possess a grammar just as we do. But it's a dry sort of kinship, located in the head rather than the heart. Rather than translate animal voices into shapes we recognise, we might discover a more acute connection via a mingled sensorium – a way of knowing the world that many animals find entirely ordinary. One way to do this might be through data sonification: transforming information we would usually encounter visually into sound.

Where sonograms offer a visual representation of sound, data sonification does the reverse. Rather than a graph like the famous hockey stick, sonified data on climate change might translate rising CO_2 as a change in pitch. On their podcast,

Loud Numbers, musicians Duncan Geere and Miriam Quick create whole pieces of music out of stories of ecological change. On 'The Natural Lottery', they transformed over a century of data recording the annual date on which ice melted on the Tanana River in Alaska into ten minutes of ice-cold techno. The music rises and falls in pitch over a motoric beat, in concert with the changing ice cover, gradually rising higher as the climate crisis begins to bite; and swirling around this is a shimmering, otherworldly wash representing the intensity of the aurora borealis, and a siren that builds in pitch as CO_2 levels rise. Another track, 'The End of the Road', is composed of data on the decline of the insect population in Denmark. Building on a baseline borrowed from the Dies Irae sequence, which has been used in Requiem Masses for hundreds of years, and a funereal chime marking the passage of each year, the track layers a chittering synth to represent the health of the invertebrate population. As the landscape is emptied of insects, the piece falls into silence.

'You look at a chart, and you don't tend to feel much,' Duncan told me, 'whereas we think that sound has a much, much stronger emotional connection with people.' Their sonifications are deeply immersive, pitching you into the sound-world of ecosystems in crisis. Music is immediate, Miriam explained; it drops us into its flow, and while the effect can be disorienting, 'it can involve you much more with what is actually unfolding in the dataset'.

'You can have a sonification so loud that it hurts,' Duncan followed up. 'But you could never have a bar chart that gets so long that it hurts. You engage with sound on a completely different sensorial level. It's multisensory.'

There does seem to be a kind of synaesthesia in their method, even though they told me that neither of them experiences a mingling of their own senses. Every project begins with the same questions: *This thing that is not necessarily audible, what does it sound like to us? What does it feel like?* The point is not fidelity in sound – their tracks don't feature any noises made by animals or endeavour to represent sounds as animals might hear them – but affinity; a sense of kinship rooted in the way that sound affects us bodily and emotionally. Their music doesn't *tell* me anything; rather (as well as manifesting as shapes shifting across the dark screen in my mind) it makes me feel with all my body what is happening in the life-worlds of Arctic caribou as early snowmelt limits their breeding season, or of birds and pollinating plants dealing with the disappearance of insects.

We humans tend to trust our eyes over all our other senses. But sound is probably the closest many of us can come to an experience of synaesthesia: even if you don't 'see' sound, we can all feel music in our bodies, whether viscerally, in the thrum of a bassline, or emotionally, in the uplift of a particular melody. What if we could encounter every story of animal change this way – from the peppered moths' fluctuating colours to the tomcod placidly swimming through waters rich in PCBs? The riot of animal improvisation in our cities could become a soundtrack, carrying what is happening in their bodies directly into ours.

What if, instead of trying to translate the animal world on our own terms, we just shut up and danced?

Sound can be a measure of health: recent studies have shown that we can diagnose how well an ecosystem is functioning by listening to it. A healthy coral reef system sounds like frying bacon; a contented hive of honeybees hums a constant A (rising to a piercing C sharp when the hive is distressed). Injured plants will emit an ultrasonic popping sound. We can even boost the health of a degraded reef through a process called acoustic enrichment: quiet reefs are less appealing to fish looking for somewhere to settle, but playing recordings of the hiss and fizz of a healthy reef can draw them back, increasing the number of species by up to 50 per cent.

Nothing in nature sounds in isolation; everything is rhymed by something else, whether it is predator and prey or pollinator and flower. Some flowers have evolved to release pollen only in response to a particular musical note. In order to reach the pollen of a spiny herb called buffalo bur, buff-tailed bumblebees will vibrate at a particular pitch – a technique known as floral sonication. The bee grips the flower's poricidal anthers, the pollen-bearing stems, with its mandibles; by contracting its flight muscles, it generates a precise tone which releases the pollen. Unless the bees produce the exact note to rhyme with the one remembered by the flower, the pollen remains locked away.

But soundscapes decay along with ecosystems – a symphony of life replaced by a vast desert of sound. As pollinators around the world have declined due to colony collapse disorder, some farmers have resorted to doing the bees' work of pollinating their orchards with tuning forks. We can do better than finding ways to replicate lost sounds, however. We can instead learn to listen, to heed the complexity and discover the kinship in animal songs.

'The world is a song,' writes the Icelandic author Halldór Laxness. Whether music or language, everything is part of the same, manifold hymn of praise in which life celebrates itself continually. The world is filling up with disfiguring noise that amputates the song of life – as toxic as any chemical contaminant – yet the song goes on. We may understand little of it, but its essence is plain: the point of the song is to be sung, without cease.

THE KINSHIP OF LANGUAGES

Orpheus was the great poet of Greek mythology, so gifted he could charm all of nature with his song. Even after he failed to bring his beloved Eurydice back to life, in the hurricane of his grief Orpheus continued singing his praise of creation. Eventually, the music enraged a group of Ciconian women. First, they ripped and tore the birds and other wild creatures held spellbound by the voice of the poet, then they turned on Orpheus himself. They pulled his body apart with their blood-stained hands, but even this did not silence him. Throughout the assault, the song went on – because, as Margaret Atwood puts it:

> To sing is either praise
> Or defiance. Praise is defiance.

The 'Barracuda Effect'

5

STRANGE MINDS

How other kinds of intelligence can help us remake our economies

In her essay 'Living Like Weasels', Annie Dillard writes of being surprised by one of the animals near her home in Virginia. Possessed by an urge to shuck off certain habits of mind, the tyranny of analytical thought and the fetters of self-consciousness, she longs to learn instead 'something of mindlessness', when suddenly mindlessness presents itself. The animal arrives – 'Weasel!' – like a shot of adrenaline; 'a muscled ribbon', all body and instinct. But for a minute, while they hold each other's gaze, she feels a connection like 'a bright blow to the brain'. The encounter lasts all of sixty seconds, and then, as unaccountably as it began, the connection breaks and the weasel is gone.

'I tell you I've been in that weasel's brain,' she says, 'and he was in mine.' But what she finds there is an unreadable blank, just 'a spray of feathers, mouse blood and bone'.

'A weasel is wild,' she surmises. 'Who knows what he thinks?'

Each animal mind is a locked box and we can only guess at what is inside. Dillard sees her weasel as an emblem of life lived according to instinct, free of self-consciousness. The weasel's watchword is necessity, she writes, whereas hers is choice. But as primatologist Frans de Waal explains, in any given species what we might call a 'mind' is the result of a complex interplay of the two: of the necessities of biology on the one hand; and choice, shaped by learning and cognition, on the other. All of this works in concert with the creature's interactions with its world, and with its unique way of perceiving that world. Necessity and choice are at the root of most animal minds, and this simple combination has produced an incredible array of highly specialised intelligences shaped by the challenges of living, say, in the latticed world of the tree canopy, or as one of a colony of millions.

Everywhere we look, nature is thinking. As early as 1900, Jagdish Chandra Bose argued that plants can sense their environments and modify their behaviour in response. In 1992, scientists discovered that wounded plants use electrical signalling to alert their neighbours to a potential threat. Some in the field of plant neurobiology argue that the root system is akin to a brain, reviving the notion of a 'root brain' first aired by Darwin in 1880. A part of the root near the tip called the transition zone is alive with electrical activity, and functions in some ways like neurotransmitters in animal brains.

In recent years, the study of non-human cognition has revealed just how diverse and widespread intelligence is. Even slime moulds have remarkable problem-solving abilities. In an experiment published in 2010, a yellow slime mould, *Physarum polycephalum*, reproduced the map of the Tokyo subway system. Placed on a map of the city, with oat flakes arranged in a pattern resembling the surrounding settlements, the slime – which grows as a single cell and lacks a central nervous system, let alone a brain – mapped the most efficient routes to each food source in mere hours.

Non-human intelligence can often look a lot like the human kind. Long-tailed macaques in Bali have not only learned to exploit the concept of trade, they've also learned the difference between high- and low-value objects (items stolen from tourists can be exchanged for food, but pilfering an electronic device like a camera gets a better return than a flip-flop, for example). Many animals use tools – orangutans, for example, have been observed fashioning gloves out of leaves to protect their hands when gathering spiny fruits. Corvids are famous tool-users, but some are also able to observe and manipulate the laws of physics. New Caledonian crows can judge the weight of different objects by watching them move in a breeze, and rooks tasked with retrieving a treat from a cup of water will use stones to raise the water level until it brings the morsel within reach. Both owls and wasps make mental maps of their territories. Archerfish, zebrafish and salamanders can count; cichlids and stingrays can perform very simple arithmetic (adding and subtracting one) as well as counting to four; honeybees understand the concept of zero, something human minds only discovered in the third century BCE and which children learn around the age of six. Pigs have been taught to play video games and rats to play hide-and-seek.

The world is full of strange minds. It is a great pity, then, that we are learning to appreciate all this just as many of these minds are being altered by us.

Changing behaviour is often a creature's first response to living on a human planet, but this behavioural change reflects an equally altered mind – how an animal acts being simply the visible result of internal processes of perception and cognition.

Electromagnetic noise is interfering with the internal compasses of migratory birds. Hermit crabs distracted by boat noise show a delay in responding to the shadow of an aerial predator. Aquatic insects are disoriented by artificial surfaces, enticed to lay their eggs on asphalt roads, cars, gravestones; anything reflective that mimics a body of water. Male jewel beetles, which gleam like brown gems, forsake females to mate with discarded beer bottles.

In the oceans, acidification can alter the neurotransmitters in the brains of some marine creatures, and thus the physical cues they use to make certain behavioural decisions, such as how to avoid predators. Coral reef damselfish raised under conditions of elevated CO_2 actually became attracted to the odour of their predator, the brightly coloured dottyback. Juvenile damselfish lose the capacity for associative learning that would teach them which fish to elude, whereas juvenile pink salmon lose the ability to identify their natal rivers. After just three days of exposure to higher levels of CO_2, sharks are less attracted by the scent of blood.

CO_2 also alters the ability of larval anemonefish to navigate via acoustic cues: typically, the young fish would be attracted by sounds of the nocturnal reef, when predators are absent, and repelled by the sounds of the more dangerous daytime reef. But high CO_2 means the cues switch; for the usually nocturnal anemonefish, reality flips and night becomes day.

Animal minds, for so long strange to us, are becoming strange to themselves. As a result, the quality of their decision-making is suffering. In this respect it's tempting to say they are becoming more like us. We know the dire consequences our way of life will bring to pass, yet carry on burning and despoiling regardless; like the anemonefish, rushing as if compelled towards the greatest danger. But the lessons of these strange minds, stretching what we understand intelligence to be, might also help bring us back to our senses.

We tend to view the mind as unified; we feel intuitively that, in any living thing thought to possess one, 'mind' is singular. But it turns out that thinking is a much stranger, more plural matter than we might suppose.

In a paper called 'The Extended Mind', published in 1998, philosophers Andy Clark and David Chalmers argue that thinking isn't confined within the skull. People often export all kinds of cognitive processes to the world around them. To illustrate their idea, they tell a story about Inga and Otto, both of whom wish to attend an exhibit at New York's Museum of Modern Art. Inga recalls that the museum is on 53rd Street and navigates her way through Manhattan accordingly; Otto, who has Alzheimer's, manages a journey he cannot hold in his mind by recording the location of the museum in a notebook he carries with him. Otto's notebook, say Clark and Chalmers, is an externalised memory. We decant our minds all the time, moving information to lists, journals, maps, road signs, and recipes, a habit that is so commonplace as to obscure the continual bleed between thinking that happens in the head and thinking that happens out in the world. The microchip, super-computing and the evolution of artificial intelligence have enlarged and intensified this extension of mind far beyond what we might once have imagined possible. Today, Otto might use a smartphone in place of his notebook: an external brain of extraordinary processing power.

Technology has extended our thinking power exponentially, but an extended mind isn't limited to humans. Mud wasps export the knowledge of how to build their nests to the nest itself. Mud-wasp nests are complex structures, consisting of a deep subterranean chamber and long tunnel that protrudes and curves several centimetres above ground, ending in a downwards-facing, bell-shaped entrance. Rather than work from an internal blueprint, which would involve keeping an image of the entire nest in mind, the wasps use the nest as a guide: when the tunnel reaches a certain height, it triggers a behaviour in the wasps to begin building downwards; if mud is added to the base, artificially shortening the tunnel, they will resume building upwards. Other insects use chemical cues to externally store memory of areas they have already searched; even slime moulds can better navigate a maze if they release a compound of glycoproteins to export memory of where they have been, like the red thread Ariadne gives Theseus to help him find his way out of the Minotaur's labyrinth.

Extending the mind into the world can make some organisms smarter. Large brains bring a lot of energy costs, but web-building spiders enhance their mental capacity by offloading a considerable portion of their cognitive function to their webs. They amplify the spider's attention, creating a wide-ranging and powerful scope which the animals can fine-tune by pulling on the threads in a particular portion of the web. In a very real sense, the web determines what the spider thinks: studies have shown that artificially tightening radial threads can force the spider to focus on areas of the web that are less profitable. The webs even tell the spider how to build them: like the mud wasps, web-building involves following a series of simple 'if this, then that' rules rather than a complete blueprint.

The spiderweb brain is an extraordinary achievement of evolution. Spiders have been building webs for at least 100 million years, possibly longer given how difficult it is for spider silk to fossilise. It is almost inconceivable, then, that in less than a century we could make some spiders dumber. Orb weavers use up to seven silks of different strengths and flexibility, the toughest of which, major ampullate (MA) silk, is strong enough to absorb the impact of bat and small-bird collisions. Exposure to the higher temperatures and humidity associated with climate breakdown alters the mechanical properties of MA, however, making it more rigid. Stiffer silk is more likely to break, and is less effective at communicating through vibration where prey has made contact with the web. One study has suggested that orb weavers constructing their webs in Sydney in the summer of 2090 will be 'critically impaired' by the changes to their silk. Spiders — which have been extending their minds into their environments for millions upon millions of years — may in the near future find the connection between brain and web is cut, as if their mind were cleaved in two.

Cognition doesn't just stray out into the world; it also goes deep inside the body. Every vertebrate organism offloads some degree of cognition, principally related to motor learning, to its spine. That same organism's microbiome plays such a powerful role in affecting its thought processes that scientists describe the gut-brain axis as a distinct cognitive system, effectively operating as a separate brain. In octopuses, mind is everywhere: of their 500 million neurons, the majority (approximately two-thirds) are in the arms. What is more, it is unclear how many minds are at work in a single animal at a given moment. Some studies suggest octopuses don't always know exactly where their arms are, as if each arm has a mind of its own.

According to Michael Levin, an expert in developmental and synthetic biology, thinking even exists at the level of individual cells. Choices get made everywhere in biology, from the most basic cellular level up to the most complex society. Every organism is a vast matrix of independent thought. 'Intelligent decision-making', he wrote in an article for *Aeon*, 'doesn't require a brain.'

Over a video call to his study in Boston, I asked Michael how it's possible to think without a brain.

'The brain has neurons, but neurons do nothing that every cell in your body doesn't also do,' he replied. Cells demonstrate all the characteristics we might associate with a thinking mind: they solve problems, create memories and express preferences. Cognition may be minimal at the cellular level, he said, but some form of it is present wherever we find life.

The origin of intelligence is in bioelectricity, the charge that powers every living cell. 'Biology has this amazing thing called gap junctions,' Michael explained, 'which are these little electrical synapses that let stuff go directly from the internal milieu of one cell to another. Imagine two cells are connected by a gap junction. The first cell gets poked with some kind of signal, and reacts to it by having a calcium flux or some other physical response that creates a memory of what just happened. That signal immediately propagates to its neighbour. Now, the second cell cannot tell that this is a false memory, something that happened to its neighbour rather than it. It's transferred almost like a mind meld.'

This is how organs and bodies are made, building acting collectives out of sensing individuals. The flow of electrical charge through pores in cellular membranes called ion channels creates a chain reaction of responses in cells, firing just like brain cells to allow them to respond, recall and even coordinate. When you have this sharing of experience, Michael said, 'it ensures cooperation, because whatever I do to you comes back to me.

'Once you share memories, it becomes impossible to distinguish *me* from *we*.'

Michael's account radically revises the picture of what counts as intelligence. It flips the narrative that places higher-order organisms at the end of a long chain of evolution; in fact, intelligence was an innovation of our earliest, unicellular ancestors. It fractures the notion of a unitary mind, and from the shards assembles a collective one. 'I am a spectacular and horrifying crowd,' was the poet Adam Dickinson's statement about the concoction of toxins he carried within himself. But even without the presence of industrial poisons, each 'I' – from the CEO rushing because she is late for an important meeting, to the weeds growing in the pavement cracks that she steps over – is many others. Every living thing is a collective intelligence.

At the cellular level, Michael said, it is often stress that binds cells together. 'In biology, stress is mediated by molecules, just like memory.' Cells leak stress molecules to their neighbours, and the neighbours can't determine if the stress is theirs or whether it comes from elsewhere. 'It's second-hand stress, but they can't tell because it's the exact same stress molecule they would produce. Stress is a mechanism to make my problems our problems. Now everyone's engaged in figuring out what to do.'

At greater scales, stress tends to do the opposite. Whole ecosystems are fragmenting under the strain we impose on them. And despite the testimony of the worst affected, the work of activists, reams of scientific data and the evidence of our own eyes and ears, the stress of living through a period of climate breakdown hasn't yet united the world in action. Vested interests and denialism continue to work like toxins in the body politic.

'There is a way of speaking which implicates your body in everything on Earth,' writes author Daisy Hildyard. It is a language of two bodies. Your first body is the one that eats, sleeps, and carries you through the world, but is limited in time and space; the second is diffuse, present in places very distant from wherever the first body happens to be. The second body is the product of choices made by the first, which conjure a vast network of complicity. It is also, Hildyard insists, a physical reality, but it slips the bounds of your skin, found instead in molecules of carbon dioxide and slivers of plastic. The second body marks our intimacy with the distant consequences of our actions.

Sometimes it can just be too much to think about this second body. Great tits nesting near a heavy metal smelter in Antwerp, where they were exposed to metals like lead and cadmium, lost interest in their wider environment. The birds showed similar levels of territorial aggression, but were markedly less curious about the world. We can all feel at different times like those incurious birds. We cut ourselves off from the rest of the world; all we can manage is to defend our little patch. 'The problem with climate reality', writes psychoanalyst Sally Weintrobe, 'is we tend to feel either too much or too little about it.' I know this feeling well: when the second body collides with the first and a molten panic rises. At other times, I try to open myself to the dreadful reality and feel weirdly impervious, as if I'm swimming in the ocean and trying to absorb the whole of it through my skin.

In *The Nerves and Their Endings*, climate activist Jessica Gaitán Johannesson observes that the act of fully confronting climate collapse must be experienced in our bodies, as something 'written into us'. It isn't knowledge that can be picked up and put down, like something in a book; facing reality means tattooing it on our skin. Where a tattoo very clearly marks the body's outer limits, however, absorbing the reality of climate breakdown means recognising that 'you do not end with your physical boundaries. Your nerves . . . seem to stretch beyond what is visibly yours.' And also, she argues: 'This is what it means to be one element in an ecosystem, and to live in a human body, to belong in its demise as well as its flourishing.'

'Where do I end and the outside world begins?' Michael asked during our conversation. It's a question that makes me look again at my body, at where it exists and how it thinks. My second body teaches me that where I end is not where the outside world begins; but it is my first body – the one I inhabit day to day – that offers a model for living through crisis. My cells don't weigh up the pros and cons of acting on shared stress; one cell absorbs the distress of its neighbour as if it were its own: this is simply what it means to be alive.

We don't just occupy animal minds; we also invade their dreams.

Nearly every living organism with a nervous system sleeps, and the need to do so may be extraordinarily old. Even jellyfish, which are little more than a nervous system in a bag of water and may have swum in Precambrian oceans for a hundred million years before the emergence of complex life, have a sleep state. Animals in every taxon – not only mammals and birds, but also fish, lizards and whales, and even insects, worms and molluscs – exhibit some form of two-state sleep, alternating between phases of low and high brain activity. Dreams occur when the sleeping brain enters a phase of high activity (known as rapid eye movement, or REM, in humans and other mammals), and there is compelling evidence that not just sleeping, but also dreaming, take place throughout the animal kingdom.

The fact that animals dream was confirmed by French neurophysiologist Michel Jouvet more than sixty years ago. In 1959,

Jouvet – who also discovered REM sleep, which he characterised as 'paradoxical sleep', when the body is passive but the brain is active – conducted an experiment to determine whether domestic cats dream. REM is accompanied by a state of atonia, when the body loses all muscle control (without this, dreamers would perpetually act out their dreams). Jouvet severed the dorsolateral part of the cats' brains, which he surmised would suppress atonia but not interfere with the high sleep state. The cats proceeded to prowl, groom themselves, stalk absent prey and even fight, hissing at empty space with backs arched and ears drawn back, all while sleeping.

Since then, research has discovered oneiric behaviour in everything from colour-changing octopus and cuttlefish, which seem to paint their dreams on their chromatic skins, to honeybees, whose antennae periodically twitch during phases of high-activity sleep. A 2001 study demonstrated that rats trained to follow a maze appear to repeat the route in their dreams. Recordings of their neural activity showed near-identical patterns between waking and sleeping: the rats were rehearsing, in dreams, each step they had taken while awake. Comparing brain activity also shows that juvenile zebra finches, which must learn complex song arrangements from older birds, sing the songs back to themselves in their sleep. The patterns of neural activity are so nearly identical, scientists have followed the dream-song note for note.

We can't know for certain what any animal dreams, but common sense would suggest they dream of their own worlds. Jouvet's cats dreamed their self-absorption; birds dream in songs of territory and sex. But it seems that increasingly we may be making incursions in these dreamscapes. In 1995, five chimpanzees at Central Washington University were observed talking in their sleep using American Sign Language, including signs such as 'coffee' that have no place in the world of wild chimpanzees. The young zebra finches whose capacity for social learning is inhibited by traffic noise also suffer declines in a range of other cognitive performances, such as motor learning, spatial memory and inhibitory control. The birds rely on their prodigious memories to learn their complex, arrhythmic songs. A poorer memory for song means weaker songs in dreams.

REM-like behaviour (eye movement; leg twitching and curling) has been found in spiders, although we don't know for sure whether or not they dream. But if they do, these dreams too may be becoming more stressful. The increase in powerful tropical storms along the Gulf and Atlantic coasts of the US and Mexico is selecting for more aggressive spiders that can better cope in a post-hurricane landscape, raising the question of whether spiders are dreaming cortisol-tinged dreams.

In some extreme instances, we are the shadow that darkens animal minds as they sleep. In one deeply distressing case, scientists exposed rats to what can only be described as torture. One group was tortured physically with electric shocks to their feet; another was tortured psychologically, forced to witness the first group's pain and distress. Not surprisingly, both groups of rats suffered trauma-induced nightmares, repeatedly startling awake as if suddenly assailed by a terrible memory.

The purpose of dreams has long been debated. Freud called them 'wish fulfilment', a notion that has a certain appeal when it comes to those animal dreams that seem to replicate their waking lives: what greater wish could a cat have than to *be* a cat, or a zebra finch simply to sing? 'I would like to live as I should,' writes Annie Dillard, 'as the weasel lives as he should.'

However, a recent theory has suggested that the true purpose of dreams is to better prepare us to solve problems in real life by offering us opportunities to play in fantasy. Neuroscientist Erik Hoel argues that dreaming evolved as a means to correct the brain's tendency to become fixated on a single way of doing things. His 'overfitted brain hypothesis' states that the kind of challenges our brains face every day are often highly similar (this would have been particularly so for our distant ancestors, whose days were shaped by addressing a fixed set of acute needs in an unchanging environment). Too much time spent learning to solve the same problems, day after day, eventually limits the brain's ability to generalise. It can respond to a certain problem very well, but can't adjust its frame of reference to answer anything new: it becomes 'overfitted' to a narrow set of circumstances.

The brain's solution is what Hoek calls a 'noise injection', in the form of dreams. Our dreamscapes are training grounds for dealing with the unexpected: not just in terms of the challenges offered (you're on a journey through a strange and delightful landscape and obstacles continually bar your progress) but also in the way the strangeness in dreams is often in the details (you find yourself in a house that you know, yet peculiar details render it bizarre). They immerse us in a heightened reality as an antidote to the deadening effect of the everyday. 'Dreams feel real while we're in them,' says Leonardo DiCaprio in Christopher Nolan's *Inception*. 'It's only when we wake up that we realize something was actually strange.' According to Hoek, that strangeness is the whole point: it gives the brain things to reckon with that it would never encounter in waking life, and in doing so helps it to deal with new situations.

Something similar may happen in the dreams of animals.

Another rat study, from 2015, found that replaying an experience in their dreams helped rats complete an unfinished task. The animals were sent into a T-shaped maze, with food in one of the smaller arms placed out of reach behind a transparent barrier. They could sense the food but not reach it. Scientists then recorded the rats' neural activity while they slept, and on waking reintroduced them to the maze – only this time, both the barrier and the food were gone. Even without the smell of it to guide them, the rats immediately sought the arm where the food had been, their neural activity exactly replicating the firing of hippocampal cells while they slept. Dreaming their way through the maze had helped them to imagine a future success they had never experienced.

Hoek suggests that the overfitted brain hypothesis also applies to artificial intelligence. Deep neural networks (DNNs) are multi-layered machine learning programs modelled on the way our brains are connected, and they face similar challenges when learning. These networks learn from datasets, and must trade the ability to remember against the ability to generalise: the more immersed a DNN is in a single dataset, the greater the risk that it will become overfitted, able to interpret the world only according to that dataset. This has particular implications for what AI is able to dream.

Unlike with animals, we can see directly into the dreams of artificial intelligence. In 2015, Google engineers designed a DNN called DeepDream with the intention of observing how the AI built a picture of the world. A DNN consists of stacked layers of artificial neurons. When the network is asked to describe an image, each layer communicates what it sees to the next layer, progressing towards an 'output' layer which provides the 'answer'. This process trains the AI to distinguish between signal and noise – to ascertain the essence of what it has been asked to identify and ignore incidental details. In this way, DNNs learn to recognise everyday objects (forks, dogs, balloons) in the real world.

The Google engineers discovered that, if they reversed the process, the neural network could not only recognise objects, it could invent what wasn't there. They could present the network with an image full of visual noise and ask it to find a particular image, say a tree. The first layer would detect whatever branching or foliate elements resembled what it understood to be 'tree' within the field of random information; each subsequent layer then built on this, leading to greater and greater flights of fancy. Mistakes were amplified: what began as a tree might begin to suggest features of a horse, or vice versa, leading to more tree- or horse-like features in each succeeding layer of detail. The images that emerged from DeepDream's 'dreams' were hallucinatory, surreal variations on familiar things. 'Horizon lines tend to get filled with towers and pagodas,' the engineers noted. 'Rocks and trees turn into buildings. Birds and insects appear in images of leaves.' Their term for the technique of making AI dream was 'inceptionism'.

The dataset on which DeepDream was trained came from ImageNet, a database founded in 2007 to 'map out the entire world of objects'. ImageNet has since labelled 14 million images according to over 20,000 categories. Significantly, the particular dataset deployed on DeepDream contained images of dogs, divided into more than 100 categories. It followed that, overfitted by its training data to see the world as dog-shaped, DeepDream built itself a mind filled with dogs, but unlike any dog you'd find on the street; its dreams were overrun with warped, oneiric mongrels. The Wikipedia entry on DeepDream is illustrated by an AI-dreamed version of the *Mona Lisa* with the tiny face of what looks like a chihuahua or a papillon, against a peacock's-feather background studded with black doggy eyes.

The pictures look strange to us, but they simply represent DeepDream's efforts to make sense of the world – to find some features it recognises amidst a cloud of white noise and make them an anchor to its reality. According to John Berger, our ancestors were drawn to make images of animals for the same reason. Animals first entered the human imagination, he writes in 'Why Look at Animals?', as 'messengers and promises', exporting the heightened possibility of the dream-world to the real one. In 2019, archaeologists discovered what may be the world's oldest painted story, in a remote cave in Sulawesi in the Indonesian archipelago. Made with vivid rust-red pigments nearly 44,000 years ago, it shows six figures surrounding a large animal with ropes and spears. The figures look human, but not quite: each has a pronounced beak, and at least one has a tail. It may be that this isn't only the earliest known hunting scene, but a far more tantalising glimpse into the minds of our distant ancestors: the first recorded myth. Animal and human mingle in this scene, like figures in a waking dream.

As they came to feature in our ancestors' stories, animals gave a kind of narrative solidity to the randomness of reality. Once they became magical, and not simply a source of meat, animals helped us impose an order to our world. Thousands of years later, they seem to have taught AI to do the same.

AI can show us impossible worlds that we might otherwise only encounter in a dream. But sometimes the world that AI reproduces is too much like the real thing. Computer scientist Joy Buolamwini coined the term 'the coded gaze' to describe how racist attitudes are encoded in AI-powered facial recognition programs. Such programs fail to recognise dark-skinned women in a third of instances, including high-profile figures such as Serena Williams and Michelle Obama. When researching facial recognition software, Buolamwini herself had to wear a white mask in order for the program to recognise her as a person.

The problem of bias in AI is endemic. In 2019, ImageNet – the same database which taught DeepDream how to dream of dogs – announced it would be removing 60,000 images from its system, after artist Trevor Paglen and AI researcher Kate Crawford revealed that its training sets were encoded with racial and gender bias. To illustrate this, Paglen and Crawford developed ImageNet Roulette, an app which invited users to upload a photo of themselves, which ImageNet would then classify. Whereas white men were typically labelled with a range of occupations ('doctor', 'scientist', 'lawyer'), women were labelled with 'slut' and dark-skinned people with 'criminal' or some form of racial slur. In 2023, a different study found that the larger the dataset, the greater the likelihood that it would reproduce prejudices.

Things get worse when AI is prompted to imagine the world for itself. When asked to produce images of people working in different professions, Stable Diffusion, a generative AI which draws on the same massive dataset (LAION) as the 2023 study, created a world that was even more divided by racial and gender bias than the real one. In the mind of Stable Diffusion, pretty much all CEOs are white men, no woman can be an engineer, and barely any women of colour work outside the service industry. The results look a lot like our world, with all its barriers and biases, but skewed towards even greater inequality.

'We are essentially projecting a single worldview out into the world,' AI researcher Sasha Luccioni told Bloomberg. This matters gravely, given the role generative AI may play in our future: some experts predict that in the coming decade up to 90 per cent of online content could be artificially generated.

Animals are not exempt, either, when it comes to AI's encoded bias. In 2019, researchers at MIT used a generative AI called a GAN (or generative adversarial network) to conjure up a menagerie of hybrid creatures from photographs of real animals. The first, the 'golden foofa', was a blend of a golden retriever and a goldfish.

'We just immediately fell in love with this creature,' Ziv Epstein, lead researcher on 'Meet the Ganimals', told me. 'What is this thing? It's delightful.'

The golden foofa has liquid black eyes and a vibrant red-gold coat almost too vivid to be real. Its wisp-like front paws look more like fins. Other ganimals are equally fantastical: a gorilla/tabby cat hybrid looked like a thickly muscled lemur. Some are comic, like the pug/armadillo that just looks like a pug wearing a supermarket turkey on its head. Others are nightmarish scrawls of black legs. Crossing a jellyfish with a bee produces an astonishing pink liquid flower.

But as the team designed different ganimals, a particular problem kept recurring. Any ganimal that included a barracuda produced an abominable centaur, monstrously half-humanoid with a fringe of mutant fins. 'The Centaur is the most harmonious creature of fantastic zoology,' writes Jorge Luis Borges in *The Book of Imaginary Beings*, but these look like chaos embodied. The researchers called the problem the Barracuda Effect.

Ziv was stumped. 'There are no humans in the training data,' he told me; or at least, there shouldn't be. The reason became clear when they examined what the AI was learning from: photograph after photograph of fishermen, proudly holding aloft line-caught barracuda. The AI could not disentangle the fish from our tendency to treat it as a trophy. Our history of violence towards other living things was like a ghost haunting the machine.

Dreams can help break us out of fixed patterns of thought and action. But beware the Barracuda Effect: unless we take care over what we feed the mind of artificial intelligence, we might find ourselves caught up in the same recurring bad dream.

From the single cell to the most complex society, we find the same patterns of perceiving and problem-solving that are the basis of cognition. Intelligence is multiple, irreducibly. But there is one thought – one enormously powerful, overfitted idea, as persistent as the worst recurring nightmare – that is driving the world into mindlessness: profit.

In the late eighteenth century, a new method of managing nature was devised in the kingdoms of Prussia and Saxony. Fiscal forestry sought to maximise the timber yield of the royal forests by reducing things to the simplest possible terms: trees that could be fed into shipbuilding and state construction were crops; species that competed with them were weeds. Forest managers planted serried ranks of a single, valued species – Norwegian spruce – with military precision. Historian James C. Scott suggests that this radically simplified view was motivated by the drive to render the forest down to a single, legible number: the total revenue the Crown could extract on an annual basis. Indeed, he notes, under fiscal forestry, 'the forest itself would not even have to be seen; it could be "read" accurately from the tables and maps in the forester's office'. Versions of fiscal or scientific forestry spread throughout Europe, the United States and the colonised world with devastating effect, scraping whole ecosystems clean of anything but the thought of profit.

Today, we're accustomed to talk of 'the market' as if it were a distinct entity, conscious and capable of making decisions. The market, we're told, can even feel, be it confidence or caution, and its verdicts carry the fate of millions. This fallacy became one of the central economic philosophies of the postwar era via the work of economist Friedrich Hayek, for whom the market represented a distributed intelligence: a vast web of agencies united by a singular intention, expressed in an unflagging will to drive competition forward. Whole economies are governed according to the demand that gross domestic product – a measure of economic output that didn't exist before the 1930s but which has become the essential index of economic and political health – should increase indefinitely.

This drive is all-consuming, literally: capitalism's defining imperative – grow! – presumes there will be infinite resources to exploit, a false dream of abundance that is devouring the basis for life on the planet. Capitalism makes the world into a vast, auto-annihilating digestive system, consuming itself and shitting excess carbon into the atmosphere and the oceans.

'What sort of animal is capitalism?' asks political philosopher Nancy Fraser in *Cannibal Capitalism*, before concluding it is an 'ouroboros', a snake consuming its own tail. But if I were to choose an animal as the icon of our self-devouring era, I'd probably choose the cane toad. Since they were brought to Queensland in 1935 to control a sugarcane-eating beetle, cane toads have spread across more than 600,000 square miles of the continent. They had little impact on cane beetles, but their effect on Australian biodiversity was devastating. There are no native toads whatsoever in Australia, never mind any that are as toxic as the cane toad, and almost everything that took a bite from one died, from domestic dogs to freshwater crocodiles. Nothing could stop them, and as they spread, their bodies became more adept at covering long distances, with longer hindlegs and an adapted gait to make more efficient use of their forelegs.

NATURE'S GENIUS

Where the first cane toads in Australia meandered, later generations have been fixated on moving forwards, relentlessly. They journey continuously through the night, abstaining from breeding on the way as if driven by a more powerful urge even than the one to reproduce, and catching insects on the move rather than waiting to ambush prey. Like the instinctive restlessness, or zugunruhe, of migrating birds, this change was genetic. The toads aren't programmed to move in a particular direction like migratory birds, but they are programmed to move inexorably in a single direction once they get going. (In this, like the cliff swallows of Capistrano, they have had considerable help from humans. Roads give cane toads a straight track to follow, often with convenient access to water, via farm dams, storm drains, stock-watering troughs, even watered lawns, and – in towns – they are lined by artificial lighting which attracts a feast of different insects.)

It isn't the fault of the cane toad that it was transported to an environment that cannot contain it. Still, its story resembles our own, in that a single thought takes over to the point where a way of life turns to ruin. The adaptation for longer legs leaves cane toads more prone to arthritis; animals at the head of the march also invest less energy in developing and maintaining their immune system. Everything is subordinate to the idea of moving forwards. Their success has been so great they have even begun to consume themselves. After eighty or so years of unchallenged success as colonisers of their new environment, cane-toad tadpoles have begun to prey on one another.

Consume; move forward; maximise profit. Whatever name we give it, this barren idea, persistent as a weed (or a cane toad), is the greatest barrier to our learning to live and think together with all of life. But before it was the bottom line in an account book, the forest was much, much more.

For the Kichwa-speaking Runa people of the Amazon, the world is Kawsak Sacha, 'the Living Forest'. *Runa* means 'person', but personhood is not a species-limited category. Their philosophy is animist: the essential unit of life, common to everything from ants to anteaters, is a 'self'. In *How Forests Think*, anthropologist Eduardo Kohn glosses Kawsak Sacha as 'a vast emergent self, made of the many selves it subsumes'. Moreover, this collection of these distinct selves – the whole living world – don't only coexist, they think together: this is, Kohn argues, 'no mere metaphor' but a statement of how life works.

Selves think together in Kawsak Sacha because, as well as representing a self, each living thing represents something to other living things, a sign that another creature can recognise. Twilight is a prompt that makes leafcutter ants, who must visit other colonies to mate – risking predation by nocturnal bats – think *safe*. A dead leaf mantis, which mimics the appearance of the leaf litter on the forest floor, relies on the failure of its predators to read signs properly: rather than *prey*, the dead leaf mantis signals *nothing to see here, keep looking*. Scaled up to the level of entire ecosystems, this becomes an intricately coordinated system of sign-making – an unimaginably dense web of attention and interpretation. Every living thing is also a living thought within a vast thinking ecosystem. Anything from a city to your gut microbiome can be counted a thinking forest, a one composed of the many.

Of course, the most essential thought that any creature thinks is itself, whether it is a dog or a damselfly. In the case of mutualistic relationships, an animal can only enter the fullest idea of itself in relationship with something else. Bobtail squid are tiny creatures somewhere between the size of your thumb and your fist. They look more like cuttlefish than squid, with eight stubby legs clustered around their mouths and a pair of swimming fins like angel's wings on either side of the mantle. They also have a bioluminescent belly, which blends their silhouette and the shadow they would otherwise cast on the sea floor with moonlight, making them literally invisible to predators. But bobtail squid aren't born with this ability: to acquire it, they need to make contact with an ocean-going bioluminescent bacterium, *Aliivibrio fischeri*. The bobtail squid's ability to fully inhabit what it means to *be* a bobtail squid is the gift of the microbe.

The bobtail squid illustrates something that is true for every living thing, us included. Relations are what make us *us*. To think the fullness of *anteater*, the animal must also think *ant*. For me to think *me*, I must also be able to think *you*. We are archipelagos, not islands.

Living signs can also be honed through millennia of evolution. The specific shape of an anteater's snout and tongue, fitted so exactly to the shape of ant tunnels, allows the animal to read the colony in a bodily way; but it is also, Kohn suggests, a sign that is 'read' and continued by subsequent generations, manifest in each body that is shaped after it. 'Biological lineages also think,' he states. By these signs, the forest makes sense to each generation of anteater.

In this forest of thinking selves, capitalism arrives as a kind of dementia. The designation of all living things as either resource or waste sunders long-evolved relationships, while chemical pollution and rising temperatures alter communities and rhythms that have been sustained for thousands of years, leaving beings that depend on thinking together isolated and marooned in darkness. But there is a way back. For the Runa, the way to the forest of living thoughts lies through dreams. Dreaming and dream interpretation are at the heart of their way of life (it is taken as read that all animals dream). The forest of living signs instructs them in how to interpret their dreams: just as the waking world can only be understood as emerging out of a pattern of living signs, it also reveals resonances between images conjured by the sleeping mind. It is, Kohn says, a matter of 'falling' into understanding; a bit like waking into a fuller life by sinking into a dream.

It was a sunny June afternoon, freshened by a silken summer breeze. The field of grass was waist-high and rippling shadows of green, red and gold flowed downhill to the estuary, where white trimmed the tops of the waves.

The field in question was part of Lauriston Farm, a new urban farm of around 100 acres in the north of Edinburgh. I was there to meet Lisa Houston, one of the directors of the agroecology cooperative which runs the farm. Agroecology is farming where balance with the natural world takes priority over profit, eschewing intensive fertilisers and ploughing in favour of regenerative methods adapted to local growing conditions, and led by communities working together. I was curious to learn how a field in Scotland might show us how to dream our way back to the living forest.

Lisa came towards me from a patch of allotments near the farm gate, smiling and cheerfully waving a pair of shears, a halo of hair blowing around her tanned face. The land had been leased from the city council by a sheep farmer for decades, and when the lease came up in 2019, she and a small group of like-minded people had applied to take it over. They'd expected to be knocked back, but to their surprise they were offered a 100-year lease. Eventually, they settled on twenty-five years. Covid delayed things, but the lease was finally signed in 2021, and by the start of the following year they were on site.

It was a huge undertaking. For decades, the farm had been

enclosed by fences and barbed wire, reserved for sheep and an occasional herd of cows. The plan was to convert the fields into a mix of community allotments and orchards, native woodland, a market garden, grazing for Highland cattle, and vital wetland habitat. Ground had to be dug over for the farm and allotments and deer-proof fencing erected; over 10,000 trees had to be planted, 8,000 of them in a single month to meet a funder's deadline. But people living nearby responded, with 500 volunteering (including me) to plant trees in the spring.

A small boy pushed a wheelbarrow past us carrying his delighted smaller sister. In the distance, someone was using a strimmer to clear more of the long grass. So far, a single strip of allotments had been planted, but eventually two and a half hectares would be set aside. 'We don't want to grow too fast though,' Lisa said. 'It's a bit of a social experiment!'

The nearest historical analogue to modern farming cooperatives like Lauriston is probably the commons. The world's oldest continually maintained commons is Cricklade North Meadow, in Wiltshire, which has existed since at least 1066. It may be even older: tradition has it that each year the meadow switches from cereal growing to animal grazing on Lammas, the first day of August, which was once part of the Roman calendar. On this small patch of land grows many of the UK's snake's head fritillary, a native wildflower with bell-shaped, gingham-chequered petals which thrives on the rich sediments laid down when the Thames floods the meadow each winter. People and plants, rivers and soils, all thinking together for centuries.

In 2012, however, the floodwaters didn't recede. That summer was the wettest on record: there was no hay harvest and no fritillaries could grow. One newspaper called it a 'colossal smothering', as if the thought of the river, supercharged by months of heavy rain, had overwhelmed all others.

Lisa led me down a path through the sea of grass towards a gate in the deer-proof fence. Inside, the grass was dissected into rectangular strips with rows of young trees. This, she said, was the beginning of their agroforestry system. Agroforestry involves planting alleys of nut- and fruit-bearing trees amongst the crops. When grown, the trees will create a microclimate, sheltering the crop; their roots will break up the soil, and their branches will host birds that eat crop-devouring insects.

At the market garden, at the far end of the deer-fenced field, we met Dav Shand. Tall and lean, with an open face and the same piercing blue eyes as Lisa, Dav managed the growing. Rows of green-leaved vegetables waved in the breeze. He fetched cups of tea and the three of us settled on garden chairs outside the shipping container that served as the farm office.

As they talked, the same ideas kept cropping up: connection, community and resilience. Everything was imagined in relation to everything else. The local communities who got involved would learn about the food system they were living in. To protect his produce from the wind, Dav had planted willows, which would also attract pollinators. They soon hoped to take delivery of half a dozen Highland cattle, natural grazers that would keep the grass in the wetland short for curlews, who like to have long lines of sight, while also spreading wildflowers and grasses in their dung. Nitrogen-fixing wildflowers seeded among the crops would feed the soil while their roots turned the earth, allowing the ecology of microorganisms, fungi and invertebrates to recover after centuries of being compacted by the narrow hooves of sheep.

There was much still to do, and years of work lay ahead.

'The connections may not all be there yet,' Dav said. But they were taking shape. It was like a mind beginning to think again – a whole host of thoughts, suppressed for centuries, flooding back.

These are ideas we have come perilously close to losing altogether. In the UK, most commons were erased by the enclosures of the eighteenth century (elsewhere, the theft of indigenous lands has had a similar effect). Where once 'unbounded freedom ruled the wandering scene', wrote the poet John Clare, whose Northamptonshire home was a paradise before enclosure arrived in 1809:

> Fence now meets fence in owners' little bounds
> of field and meadow large as garden grounds
> in little parcels little minds to please.

In 1968, Garrett Hardin's influential article 'The Tragedy of the Commons' seemed to sound the commons' death knell. For Hardin, the commons were doomed by the inevitability of human failings. Left unregulated, we cannot help ourselves, gutting and exhausting nature's gifts, he pronounced with patrician regret. It was a tragedy in the Shakespearean sense, and of like proportions. 'Ruin is the destination toward which all men rush,' Hardin declaimed. The only safeguard, he argued, was the oversight of private ownership.

Hardin saw our self-devouring proclivity as a regrettable but indelible strain in human nature. But his judgement rested on a fundamental misunderstanding. The commons are not resources to be managed, but an expression of a relationship. For historian of the commons Peter Linebaugh, 'commons' is a verb before it is a noun: *commoning* brings communities together for the benefit of all. Privatisation severs relationships, building walls and erecting barriers; but the thing about relationships is that, when allowed to flourish, they are inexhaustible.

More recently, the idea of the commons has had a resurgence. From community-managed forests in India to global knowledge exchanges like Wikipedia, the idea that resources can be managed for the common good – that relationship trumps resource – is transforming public life. In 2009, Elinor Ostrom became the first woman to win a Nobel Prize in Economics, for work that argued resources should be managed locally and democratically, with solutions fitted to their circumstances, by the people that shared those circumstances; in other words, through cultivating the commons.

Critics of the commons tend to compare it to planting a few seeds when the whole forest is burning. They argue that small-scale, local projects are not up to the challenge of a planetary crisis. But as activists David Bollier and Silke Helfrich suggest in *Free, Fair and Alive*, the commons is not simply a means to make everyday life better; it is also a way to reimagine our whole future together. One of the greatest barriers to the commons is in language, and the binary perspective entrenched by opposing *I* and *we*. What we need instead, they propose, is a notion of 'Nested-I' – an understanding that we all contain multitudes, and that we should carry this sense of abundance into our social and ecological relationships. Like the cellular decision-making Michael Levin described, *I* and *we* blend into one another; just as collections of individual cells become organs, and organs come together to form organisms, embracing a kind of multicellular thinking in this way could be the beginning of a wholesale reimagining of *everything*. There is no blueprint for the commons, beyond a few core principles (autonomy, democracy, and a commitment to the common good being crucial); rather, Bollier and Helfrich argue, commons are 'living systems that evolve', with ideas spreading and adapting to suit different times and places. The power to think otherwise, to initiate change, lies with the cells as well as the head.

Such a reimagining could help us reinvent how we organise our economies, our infrastructures, even our politics. Growth, defined as GDP, has been the undisputed foundation of economic thinking for generations. But the notion that we can continue to grow economies indefinitely, regardless of the carrying capacity of the planet, is a fantasy. Not a single country on Earth currently meets the basic needs of its citizens in a sustainable manner. Some economists argue that relinquishing GDP is the key to unlocking a planet-wide transformation: replacing it with a focus on securing human well-being, and recognising that this well-being is inextricable from the health of the natural world, might not only lead to rapid decarbonisation but also arrest the decline of industrialised economies that are increasingly struggling to meet their growth targets. Degrowth involves scaling down our dependence on finite resources, recognising that they are finite and instead organising society around what can be regenerated, shared and managed in the interests of the many. It may sound like a reaction against progress, but the emphasis is on changing the flow of energy through our economies, rather than on shrinking them. One study has estimated that by combining renewable energy technologies with a drastic reduction in demand for finite resources, we could reduce global energy consumption to the levels of the 1960s by 2050 – 60 per cent lower than they are today – even with a population almost three times as large, and without a decline in living standards. As economic anthropologist Jason Hickel, a key proponent of degrowth, argues, easing demand in the global north could represent a decolonisation of the global south, freeing economies that have been restricted to the role of exporters of raw materials so they are able to innovate on their own terms.

Underlying all this is a fundamental change in perspective, an alteration of mind. Degrowth would involve tapping into the latent possibility in the commons. Visualising decisions we make locally in the context of a far larger network of organisms and ecosystems that think in concert would reorient our politics away from border-riven self-interest. Our minds have become overfitted to a way of doing things drilled into us by centuries of exploitation of natural abundance in the name of profit. But the dream of the commons can help us break out of the old way of doing things.

I asked Dav and Lisa what idea guided them. 'To me, it's about the opportunity to find solidarity across dividing lines,' Dav volunteered.

'Abundance,' Lisa said after a pause. 'It's about being able to show abundance.'

We can't grow a vision of abundance if we limit it to ourselves. Nature is at the heart of the commons. Like the Runa's living forest, commons create a plurality of selves rather than mere economic value.

A few weeks later I returned to the farm to meet Leonie Alexander, an ecologist advising how biodiversity on the farm can be restored. When she first visited, she said, 'it was apparent to me that this site is just debt-ridden'. Centuries of exploitative land use had cleared the forest, reshaped its topography, ploughed, reseeded and polluted its soil, and most of all, evicted the majority of the species that once shared it.

But the land was beginning to think again. Without sheep grazing everything down to an even sward, cocksfoot grass – a tall, tussocky plant whose roots, Leonie said, go about as deep below ground as the plant stands above it – was spreading across the farm. The tough roots were breaking up the compacted soil, releasing nutrients and allowing wildflowers to grow, thus returning a whole host of commensal relations to the farm. Hoverflies have mouthparts that are specifically fitted to feed from certain flowers, such as oxeye daisies, common daisies, dandelions, chicory and field scabious; the hoverflies were attracting bats and birds. A chain of living thoughts was beginning to form.

'This wants to be woodland,' Leonie told me. 'This site, left to its own devices, would say, "I want to be a wood."'

The competing demands of people and animals meant this wasn't possible, however. The farm had been organised accordingly into thirds: a third for growing food, a third for recreation and a third for wildlife, and this could create tension. People needed a place to walk their dogs, which sometimes disturbed nesting birds; wintering curlews needed a place to rest on their migrations to and from Scandinavia. The farm sat under the flightpath for the nearby city airport, and the noise of passing planes sometimes made it difficult for us to hear one another; but they also made it impossible to forget the wide world beyond, and all its contradictions.

Leonie said that their intention was to create large scrapes – shallow pools that dry up in summer – that would provide habitat for wading birds especially as sea level rise was likely to erode their other habitats along much of the rest of the shoreline. But the airport was concerned about bird strikes. Attract too many birds to the new wetland, and there was a risk that some could get caught in the engines of jets. 'Whose space is it,' Leonie laughed, gesturing above our heads, 'curlews' or airplanes'?'

The compromise was to create a series of smaller scrapes,

seven in all, which would satisfy the airport and still make space for birds. Even out of this accommodation, a new sequence of living thoughts was already taking shape: caddis flies were performing their mass mating dances in the ponds, drawing other migratory birds like swallows and house martins to make the north field a waypoint on their journeys. Leonie said there were also dunnocks and starlings in the fields, even woodcocks and snipe, as well as summer migrants like chiffchaffs, whitethroats and willow warblers. 'And it'll only get better as the vegetation diversifies and there are more resources for everything.'

Martins and swallows wove in the air above our heads; higher, a descending jet growled and whined.

This field couldn't go back to what it was, Leonie said once the jet had passed. What they had to do was coax it into the shape of a new thought – one that was good for all.

Cognitive dissonance is an almost inevitable consequence of life under late capitalism. We depend on a system that is hostile to life, to keep us alive. We read daily about collapse, and continue scrolling. We are the jewel beetle disoriented by the gleam of our own trash, and the omnivorous cane toad consumed by the thought of a new frontier. We are the damselfish, drawn inexorably towards the greatest danger.

But we don't need to be. Our way back to the living forest is through the commons, and the dream of an abundance of living thoughts.

Many species have learned how to think and make good decisions together. A study which used high-resolution GPS to track bighorn sheep and moose in the western United States found that the animals used crowd-sourcing to find and follow the most fertile grazing areas – 'riding the green wave' by sharing knowledge accrued by the herd over decades. Another study found that wild olive baboons in Kenya base important decisions on the feelings of the group, despite the fact that baboon troops are dominated by high-ranking males. Some species give greater weight to the input of those with the greatest needs. Herds of plains zebra give precedence to the suggestions of lactating females regarding the best places to find food and water.

The world is full of life that thinks and works together. Honeybee swarms decide which nest site to colonise via a dance-off between scout bees, who each return to the hive with an alternate proposal. Once a single site has been selected, the entire hive comes together in frenzied dance just before the swarm takes off. Flocks of birds and shoals of fish perform extraordinary choreographies that are wholly spontaneous, and yet no animal is ever left behind.

Despite their complexity, murmurations – when vast clouds of starlings move and pulse in time like the realisation of a dream – are disarmingly uncomplicated. Computer simulations have revealed that to perform their aerial displays, birds must follow just two simple rules. First, each bird sticks with a small group of five or six others, coordinating its movements with its neighbours – each slight tilt of wing or handbrake turn in direction is a small decision that ripples through the whole. Next, they need to orient themselves within the whole flock. Technologist Mark Kraus writes that starlings will 'aim for the optimal light density within the murmuration; like a light meter measuring for the right exposure, the starling rolls its wings to find that perfect place where the darkness and light are just right'.

Imagine what it must be like to be inside a murmuration; the white noise of thousands of beating wings darkening the horizon. It would be overwhelming, surely, to be always so close to the edge of collapse. The birds keep up the dance by finding a point of balance, where the dark allows just enough light through so they can see that they aren't alone, that they are moving in the embrace of an incredible abundance, and that everything around them is thinking in concert.

The Silent Room at the Future Library (2014)

6

WILD CLOCKS

How rethinking time can help us choose a better future

One summer day in Paris in 1729, a French astronomer observed a plant growing on his windowsill. Jean-Jacques d'Ortous de Mairan noticed that the plant – *Mimosa pudica*, sometimes known as the sensitive plant because of its tendency to recoil when touched – would stretch its leaves towards the sun in the morning and fold them away when evening fell, as he himself was preparing to gaze into the night sky. Even when Mairan shut the plant away in a cupboard it kept its rhythm; despite being enclosed in darkness, the leaves would unfurl in time with the rising sun. 'The Sensitive-plant senses the Sun without ever seeing it,' he recorded in an account of his experiment.

What he had discovered was not evidence of a hidden heliosensitivity, however, but something perhaps even more extraordinary: a clock, buried in the plant's cells, quietly keeping its own time independently of the rising and setting of the sun.

Biological clocks are found in every organism, from gut bacteria, fungi and plants, to fish, animals and ourselves. These clocks are 'endogenous', meaning they are embedded deep within the body and tick away regardless of where the organism happens to be. But they are also highly attuned to different environmental factors. Almost all plants exposed to sunlight become entrained to a 24-hour, circadian cycle (one exception is *Mimosa pudica*, which folds its leaves every twenty-two or twenty-three hours). Sensitivity to temperature and light are the main cues, but there are other factors too: some marine animals align their spawning events with the lunar calendar, and shorebirds time their foraging with the tides. Coordinating endogenous rhythms with environmental signals will produce an 'internal clock time' that is distinctive for each organism, cued to every key life event, from when to migrate to when to flower or breed.

The revolutionary implications of Mairan's observations were neither fully appreciated nor confirmed for several centuries. Instead, it was another clock that changed the world. The year after Mairan's experiment, a 37-year-old Lincolnshire clockmaker called John Harrison arrived in London. He brought with him a manuscript which he hoped contained the answer to one of the most pressing scientific problems of the day.

The story of Harrison's sea-watch is well known to anyone who has read Dava Sobel's *Longitude*: because of the rotation of the planet, to determine their position a sailor needed to compare the time in the place they were with the time at a point of known longitude (or 'prime meridian') and calculate the difference. No one had been able to build a clock that could keep time at sea, however, where changes in temperature and the motion of the waves caused clocks to vary wildly. Countless ships were lost when they failed to tell the time correctly. Even Sir Isaac Newton, the leading scientist of the age, thought longitude was an intractable problem. Harrison's design used a pair of interlinked brass dumbbells in place of a pendulum, to compensate for the movement of the ocean. In the ensuing decades, he conceived a series of marine clocks, each one more accurate than the last. His fourth, a pocket-sized sea-watch called H4, lost less than a minute crossing the Atlantic in 1761. The longitude problem was solved.

Harrison succeeded by isolating time. Like Mairan's plants, opening and closing in the dark of a cupboard, Harrison's clocks were not subject to any external cues. Operating as if in a vacuum, they were immune to changes in heat and motion; wherever they travelled, time remained tied to the Admiralty clocks in Whitehall. But Harrison's discovery took time out of the body and made it a commodity. Time, now fixed, could be sliced into shifts by industrialists and exported by nations with imperial ambitions. In 1851, the prime meridian was established at Greenwich in London, an immovable point from where the colonisation of distant lands could be coordinated. The manipulation of time was the foundation on which we made a human planet. Without Harrison's marine clock, the Industrial Revolution, European colonialism, globalised trade, the devastation of ecosystems and the mass introduction of invasive species across oceans would not have been possible.

In the early twentieth century, circadian rhythms were observed in a wide range of animals, including bees and fruit flies, as well as zoo animals deprived of natural light. The advent of space flight, when astronauts were removed from the influence of all possible earthly cues, confirmed that the same sense of time is encoded in the human body. Life keeps its own rhythm. Time is felt in the body as much as told by the clock. By then, however, the only clock that mattered was the one on the wall.

Barbara's email was timestamped 07.17:

NATURE'S GENIUS

Hi David,

A classic, regarding our definitions of time. I noticed that you specified in GMT. But since you are on summer savings time that would be 11 am local time and 12 in Switzerland. Right?

Talk soon.

Barbara Helm is a chronobiologist – an expert in phenology, the timing of life cycles – at the Swiss Ornithological Institute. The Swiss are famous for their mechanical timekeepers, but Barbara researches 'wild clocks': the traits that enable living things to coordinate their way of life with the world around them. I had arranged a video call with Barbara for that morning so I could learn about how different species keep time, but I had forgotten about the shift to daylight savings. Pegged to the prime meridian, my sense of time was stuck in the wrong season.

To understand a crisis, we need to know what time it is; or rather, what *times*. The ecological crisis we face is also a chronoclasm – a sudden, sometimes violent collision of different orders of time, where temporal patterns break down and confusion reigns over exactly what time it is. Modern life suggests to us that time lives outside the body, running independently of the living world like Harrison's sea-watch. But as the familiar rhythms of 24-hour news and five-year election cycles unfold, a host of other measures and cadences, running at many different speeds, are playing out. The climate is changing so rapidly that many species cannot evolve quickly enough to keep up; many wild clocks are losing time, running too fast or too slow; some are winding down to a fatal stop. Our insistence on telling the time by the clock on the wall keeps us from noticing the many other ways we could coordinate our time with the world. But adopting wild clocks might help us to know what time it really is.

When she appeared on my screen a few hours later, Barbara greeted me with a broad grin and waved away my apology. There was a painting of a long-legged wading bird on the wall behind her. Her grin broke out again when I asked her what it was.

'It's a water rail,' she said. 'My home is basically full of water rail!'

Barbara's research is focused on the bodily timekeeping of birds – 'biological witnesses', she calls them, to the many different ways that life keeps time. There are three components in any organism's biological clock, she explained, which make up its 'chronotype' or temporal phenotype: an expression of time totally distinct to that creature. First there is 'body time', like the circadian rhythms Mairan observed in his sensitive plants, which is endogenous (embedded in its tissues). Although Mairan's plants were not dependent on the sun, for most living things body time is coordinated with a second component: what chronobiologists call zeitgebers or 'time-givers' – environment factors such as daylight or temperature that modulate an animal or plant's body time to a specific tempo. Their role is to synchronise the animal or plant with the world around them. Arctic-breeding birds, for instance, may coordinate the memory of when snow melted in past years with changes in day length and observations of the behaviour of other migrating birds, to decide when to begin the long journey north.

'When people first observed these rhythmic behaviours and activities, their main image was the clock on the wall,' Barbara said. 'They thought there should be just one and it should be located in the brain somewhere. They actually used what I think was quite horrific terminology for it, they called it "the master clock".'

In fact, keeping time is a far more intricate process. An animal might have different clocks in different organs, all keeping different time and serving separate purposes: one in the skin to coordinate temperature, another in the liver to regulate feeding cues. The same is true for insects: butterflies navigate by the sun, but they also employ 'antennae clocks' to compensate for changes in latitude as they migrate. Each cell carries its own timekeeper, precisely tuned to a particular function.

'And so that brought up questions,' Barbara went on. 'How could you get meaningful time information from literally billions of clocks?!'

The answer is by coordination. Time is made in the body, but it is also made by different bodies working together. The third element in a chronotype is the interaction between organisms.

'For example,' she said, 'in the beehive they make time together.' Honeybees have a social clock determined by the division of labour. Nurse bees, which remain in the hive to care for the brood, keep to a consistent rhythm that ticks away in their genes, whereas foraging bees manage a complex relationship between circadian rhythms, an oscillating clock gene, and what Barbara calls 'flower time', the separate chronotype of the plants to which the bees are drawn. Making time is a complicated negotiation between species; and also within species. A long-term study of plants in the Arctic tundra has revealed that separate timekeeping mechanisms have evolved in different parts of a single plant species, and flowering has co-evolved with the activity of pollinators while leafing has co-evolved with herbivores. These reproductive and vegetative clocks, each cued to the time of a distinct insect or animal, run at different speeds in the same organism.

Wild clocks don't tick with the steady pulse of a Swiss watch, they *swing*. Assembled together, they form a vast polyrhythmic fugue, an impossibly complex arrangement of syncopated beats and pulses, tempos layered upon tempos, in a rich, immersive cross-rhythm that drives life forwards day by day, year by year, season to season.

With so many intricate timings involved in producing a biological clock, Barbara explained, the question becomes 'what time is the relevant time for an animal?'

The advantage of this complexity is that some creatures, if they possess enough plasticity, can find new rhythms within the overall score – amending the timing of key life events as temperatures warm and the seasons slide out of shape. Sometimes making time together can even lead to the emergence of new species. Ralph Waldo Emerson called the apple America's 'national fruit', but it was in fact brought to North America by European settlers. One species introduced in the early nineteenth century set off a wave of changes called a trophic cascade: in just 150 years, a native species of fruit fly called the apple maggot and its parasitic wasp have evolved into several genetically distinct varieties. Before colonisation, an apple maggot's natural host was the hawthorn but, like Emerson and millions of other Americans, some of them developed a taste for non-native apples – so much so that two distinct types of maggot emerged, a hawthorn- and an apple-feeding kind, which no longer mix or breed. This divergence set in train a parallel shift in the parasitic wasps that lay their eggs in apple maggot larvae, which also split into distinct lineages.

The driver of this divergence was not in either the maggots

or the wasps themselves, but in the clocks embedded in the DNA of hawthorn and apple trees. (Behind this, we might also detect the steady tick of Harrison's sea-watch, which made it possible to navigate ocean crossings accurately.) Each host plant fruits at a different time of year; dependent as they are on these arboreal timepieces, the maggots who developed a preference for apple and the wasps who preferred their larvae entered into a wholly different chronotype than their hawthorn-loving cousins. In telling the time together, apple trees, apple maggots and their parasitic wasps – with the help of European colonists – also conjured a new species.

Many wild clocks are struggling to keep time, however. The beat does not fall where it should; syncopation becomes dissonance. The Central England Temperature series, the longest-running meteorological record in the world, with data stretching back to the mid-seventeenth century, has shown that the warming climate has caused significant disruptions in plant phenology. The seasons no longer mean what they once did. One survey of over 200 northern hemisphere species found that rising temperatures are bringing spring phenology forward by 2.8 days per decade. But crucially, different species are responding at different speeds. Amphibian phenology is changing twice as fast as that of trees, birds and butterflies, and nearly eight times the rate of change in herbs, grasses and shrubs. Biological clocks that have evolved an exact synchronisation over millions of years are starting to fall out of sync, creating what chronobiologists call phenological mismatches: misalignments in time between predators and prey, herbivores and plants, or flowers and pollinators, which can have catastrophic results.

In Australia, mountain pygmy possums are leaving hibernation before the emergence of their preferred prey, the bogong moth, risking starvation. Great tits are laying their eggs after the caterpillars that nestlings feed on have reached peak biomass, limiting their access to a vital food source. One study warns that the duration of phytoplankton blooms could be shortened if the oceans continue to warm, introducing a calamitous mismatch at the very base of the marine food chain.

Food isn't the only concern. In Canada, reduced snowfall is exposing snowshoe hares to predators, as their seasonal colour moult from brown to white no longer provides effective camouflage. Plants are losing touch with their pollinators: warm springs in Japan have led to the earlier flowering of spring-ephemeral plants relative to their pollinating bees. The composition of entire ecosystems is changing in accordance with these new rhythms. Around the world, animals and plants are following rising temperatures to higher latitudes at a rate of more than ten miles per decade.

Timing is especially sensitive for long-distance migrants. A timely migration depends on an intricate coordination of birds' internal clocks and environmental conditions. Migrant birds have evolved an ingenious timekeeping solution to meet these demands. In addition to their circadian clocks, which measure out daily activities, migrants possess a nocturnal clock that is tuned to the bird's particular zugunruhe – its innate migratory restlessness. The impulse is extraordinarily powerful: when the time comes to travel, even confined birds will hop around their cages, brimming with the need to move in the direction of their breeding grounds. The impulse is so hardwired, not even mountains get in the way. Bar-headed geese began travelling from their winter grounds in Assam to nest on the Central Asian plains before the Himalayas rose from the earth, 50 million years ago. As the mountains grew, the urging in their blood kept them fixed to the same timing and the same route, until today their descendants have evolved to travel a thousand miles over peaks 6,000 metres above sea level.

These mechanisms also need to compensate for the fact that birds who travel very long distances have no way of knowing exactly what they'll find when they arrive. Red knots live on the German and Danish coast of the Wadden Sea but breed 3,000 miles away, in northern Canada and Greenland. To arrive on time to lay their eggs in early June, they must leave in the first week of May, stopping in Iceland to refuel, with preparations for the journey beginning weeks earlier. In the absence of an avian weather service that can discern the date when the Canadian spring will arrive two months into the future, red knots have to rely on precise internal timekeeping, fine-tuned over millennia.

Climate breakdown is throwing many of these delicately balanced arrangements into disorder. Some birds are altering their laying dates but not the timing of their migration, shortening their breeding season. As ever, plasticity is key. Others find that new routes open up: warming has allowed some blackcaps to migrate from Africa to the UK rather than the Iberian Peninsula. Long-distance migrants like red knots can modify their timing on the wing if they sense changes in air temperature, slowing their journeys or postponing laying, but birds that rely heavily on daylight to time their migration and lack sensitivity to temperature are especially vulnerable.

The other major factor putting biological clocks out of time is urbanisation. Sensitivity to photoperiod, or changes in light as the seasons wax and wane, has been the strongest and most reliable time-giver for most of evolutionary history. But the invention of artificial light created a rival to the sun. Electric light now cues songbirds to sing at night. Redshanks observed foraging at night in an intertidal pool lit by a petrochemical plant changed their behaviour to use visual rather than tactile means to find food. Skyglow can disrupt insect migration and alters the circadian rhythms of pollinators (pesticides that target their nervous systems may also disrupt insect's internal clocks). And where there is light, there is usually heat. The urban heat island effect is altering how other species tell the time. City warmth means urban blackbirds reach sexual maturity a month earlier than rural birds. Through a combination of warming and artificial light, urban trees in the UK, Europe and China produce leaves between four and seventeen days before rural trees of the same species. The earlier greening can even be seen from space.

For all that our clocks appear to keep us apart from this, humans have time-givers too. 'We carry time in our bodies,' Barbara told me. It's why we feel jet lag. Up to three people in a hundred will experience seasonal affective disorder. 'Humans are also seasonal animals,' she went on. 'If you look at churchyards and historical birth records, there used to be very strong birth seasonality in humans. The Industrial Revolution dampened this rhythmicity but it does not go away entirely.' The rhythm of human reproduction synchronises with the seasons, with most American babies born during the long sunny days of August and September, while the fewest arrive in the short, dark days of February.

Countless rhythms pulse in every organism. Some are more pliable than others. As the planet accelerates through dramatic changes in climate, some species will adjust their clocks; others will fail to keep time altogether. I asked Barbara if we might be better off learning to tell the time by wild clocks.

Yes, she said, but the time to do so seemed short. 'So much is getting destroyed, but on the other hand, there is also this creative, transformative potential there. I go back and forth between seeing this as a broken world and an inventive one.'

Clock-time as we know it began with a light swaying from side to side. Sometime towards the end of the sixteenth century, Galileo Galilei noticed that oil lamps hung from the ceiling of the cathedral in Pisa would swing back and forth with a regular rhythm. Intrigued, he timed the lamp's swing – measuring the effects of initiating the swing at different angles, varying the length of the rope, or adding extra weight. He discovered that pendulums are very nearly isochronous: regardless of amplitude – whether the lamp was launched at shoulder height with great vigour, or nudged lightly with a finger – it would swing with the same period ('period' being the term for a complete oscillation away from and back to the point of equilibrium). A pendulum will keep its own time, at least until the energy put into it dissipates. Galileo's discovery was the foundation on which, half a century later, Dutch mathematician Christiaan Huygens built the first pendulum clocks, machines so accurate they would lose only fifteen seconds each day.

What Galileo realised was that time, which for all human history had been determined by the rhythms and forces of the natural world, could be regulated. Prior to the mechanisation of clocks, the duration of a day or an hour depended on the time of year, lengthening in summer and contracting in winter according to the Earth's orbit around the sun. Seasonal rhythms comforted and consoled, offering glimpses of the greater wheels within which our small lives turn. In Shakespeare's twelfth sonnet, 'the clock that tells the time' is found in 'the violet past prime' and 'summer's green all girded up in sheaves'. Time was a matter of shared geography: every town and village kept its own local time, aligned with the position of the sun as seen from the public square. The pendulum made it possible to capture time, wrestling it from the sun and fixing it to a regular, unchanging, universal rhythm. In time, the rest of the world fell into step with the invariable swing of the pendulum.

In an industrial society, time is a commodity like everything else. The railways require a single, standardised time in order to run punctually: in England, local times were supplanted by standardised time in 1848, when the railway schedules were oriented to GMT; in the US, four time zones replaced the multitude of local and regional time zones in 1883. Karl Marx proposed that, after gunpowder, the compass and the printing press, the clock and the mill were the most essential devices of industrialised society.

The way this story tends to be told suggests that clock-time rose to prominence unassailed, as if the clock were a new sun eclipsing all other ways of keeping time. Yet in America, it was still possible for local legislators to opt to keep local time until 1967, only two years before the moon landings. In fact, going all the way back to Galileo's discovery, other rhythms proved essential to establishing order over time. In order to measure the pendulum's swing, Galileo needed a way to mark time accurately; yet no such machine existed. To capture time, he used his own pulse. We might say that all clocks made since then have been set to the heartbeat of Galileo.

For most of us, clock-time *is* time, a perception that masks the many, different ways that time can be experienced. But if this notion flattens what we understand time to be, caging it in mechanisms designed to impose order and regularity, it also limits our understanding of what is a clock. For my friend the philosopher of time Michelle Bastian, clocks are much more than the clock on the wall. Those we rely on day-to-day are political devices, designed as much to organise behaviour as to measure time. But our habitual ways of telling the time have their limits. 'When we look at a clock or calendar we can see fairly quickly whether we are becoming out-of-sync with some worlds, but not with others,' Michelle writes. This time-blinkeredness can have devastating consequences: a conventional clock can tell me whether we are late for work, she notes, but 'it cannot tell me whether it is too late to mitigate runaway climate breakdown'.

Michelle argues that anything can be a clock, if we consciously coordinate ourselves with it. Despite the ascendence of standardised time, there is a long tradition of adopting other living things as our time-givers. Barbara told me of agricultural calendars she had seen in Indonesia that marked which crops to grow according to the arrival of different migratory birds, aligning human, bird and plant time. In 1751, the Swedish taxonomist Carl Linneaus described the *horologium florae*, a wheel illustrating the hour when different 'solar flowers' open and close, by which 'without the help of a clock, or seeing the sun, [one] might know the time of the day'.

We have naturalised clock-time, as if what a mechanical or digital clock tells us is the only time it could possibly be; yet fundamentally, a clock is what we make it. 'Clocks are artefacts that are designed,' Michelle states. 'They can therefore be redesigned.'

Perhaps no one understood this better than Henry David Thoreau. Thoreau was acutely aware of the encroachment of standardised time in his day. 'The startings and arrivals of the [train] cars are now the epochs in the village day,' he observed in *Walden*. While many of his neighbours chose to coordinate with modern machines (the trains appeared 'with such regularity and precision . . . that the farmers set their clocks by them'), for Thoreau, time was made together with the entire living world. The journal he began at Walden Pond in 1850 runs to over 1,800 pages of phenological observations. On 21 August 1851, he witnessed the blossoming of the delicate, star-shaped indigo flowers of the blue vervain in a coordinated sequence of successive circles of buds rising up the plant from stem to tip. 'This triumphant blossoming circle travels upward,' he noted, 'driving the remaining buds off into space.' He recorded the first day on which he noticed the buds breaking open, 16 July, and their subsequent flowering, and remarked: 'It is very pleasant to measure the progress of the season by this and similar clocks.'

This kind of close observation became an obsession. In a later entry, from December 1856, he claims to have 'often visited a particular plant four or five miles distant, half a dozen times within a fortnight, that I might know exactly when it opened'. In his eulogy for Thoreau, Ralph Waldo Emerson recalled how he once declared that, if he should wake up from a trance in the middle of a swamp, with no memory of when or how he got there, he would be able to tell 'by the plants what time of the year it was within two days'. As a result of years of such close and compulsive attention to the rhythms of the natural world, Thoreau came to see its times as indivisible from his own. 'These regular phenomena of the seasons get at last to be . . . simply and plainly phenomena or phases of my life,' he wrote in October 1857. 'The seasons and all their changes are in me.'

In his final years, Thoreau began transforming this wealth of phenological information into a device for keeping time with the seasons. Thoreau's 'Kalendar' compiled every reference recorded in his journals into a compendium of observations about the turning seasons; significantly, he included not just when the ice on Walden Pond formed and shrank or the day when a particular flower bloomed, but also the facts of his own body in time. He wove the days when he began to 'wear one coat commonly' or 'sit below without fire commonly' into the tapestry of changing weather, blossoming flowers, falling leaves and animal migrations. Time, for Thoreau, was first and foremost something he felt. The many wild clocks of Walden Pond ticked in his body.

Thoreau shows how coordinating with wild clocks can inaugurate a profound change in our sense of time. Michelle argues that time made together with other living things would liberate us from the constraints – socially and conceptually – of standardised clock-time. Of course, achieving this is a different matter today than it was for Thoreau. The wild clocks he knew no longer keep the same time as they did formerly. Plants around Walden Pond flower on average seven days earlier than when Thoreau recorded his phenological observations in the mid-nineteenth century. Yet, because of this difference, wild clocks can help us keep track of how time slips and wobbles through the coming decades of climate breakdown. The world's longest-running phenological study, which has coordinated data on the flowering of common lilacs across the Unites States, has measured the rise of anthropogenic carbon in the atmosphere since the 1950s. The UN's Intergovernmental Panel on Climate Change, which assimilates the latest climate data from around the world, has stated that changes in phenology can be an essential tool in tracking climate breakdown.

One study has even suggested that wild clocks in songbirds can help us to predict the likelihood of hurricanes. The veery is a thrush that migrates from North to South America, passing through the Gulf of Mexico during the peak of the Caribbean storm season. The birds spend less time on the nest in years with more extreme weather, allowing them to avoid the worst of the storms on their journey south. Observing their breeding behaviour could help us to predict how severe the coming Atlantic storm season will be.

Of course, a veery clock won't replace the predictive powers of meteorological science. But as climate breakdown whips up ever more destructive tropical storms, telling the time with these tiny birds, who brave the fury of Atlantic hurricanes on journeys lasting tens of thousands of miles, would help us to appreciate what it means to make time together with other species. By synchronising our time, we can ease the pressures forcing some creatures to live out of step with the world.

Time lives in the body, not as the tick of the clock, but as a pulse in the blood. It is a thought, buried deep in nerve, leaf and gene. Wild clocks can help us reckon with a changing world and recalibrate our sense of time, fostering rhythms by which all life can flourish.

Once again, my sense of time had let me down.

I thought I had plenty of time to spare. I'd woken early, and set off through the centre of Oslo at a leisurely pace, keeping an eye out for somewhere to buy coffee and looking forward to reading my notes in the morning sunshine before meeting Ane outside the train station. But the clock on Oslo Cathedral told me otherwise. My heart leaped unpleasantly as I looked up and saw the hands were not arranged in the neat ninety degrees of 9 a.m. as I had expected, but narrowed in the acute angle of 10 a.m.

I had relied on my phone to update time zones when I arrived the day before, but that particular setting, I later learned, was switched off. I was an hour out of step with the city around me, and more importantly, I was late.

I'd travelled to Norway to visit a forest that is also a library of books that will grow – unread – for a hundred years. Future

Library is the invention of Scottish artist Katie Paterson. Every year, a writer contributes a work to be held in trust until the year 2114, when an anthology of all contributions will be printed on paper from trees specially planted in Nordmarka, a vast forest of pine, birch and spruce outside Oslo. Each new deposit is marked by a summer handover ceremony in the grove.

The first writer to add their work, in 2014, was Margaret Atwood. 'How strange it is,' she wrote at the time, 'to think of my own voice – silent by then for a long time – suddenly being awakened, after a hundred years.' David Mitchell, the second writer to take part, called Future Library 'a vote of confidence in the future'; for the fifth, Han Kang, it was like 'a century-long prayer'. Future Library is an exercise in temporal fine-tuning, calibrating our fraught and anxious present with a future about which no one can be certain.

Even a project that adopts such a long view was not immune to the disruptions of the Covid-19 pandemic, however. The contributions of the three most recently selected writers – Karl Ove Knausgaard, Ocean Vuong and Tsitsi Dangarembga – were delayed as time around the world suddenly aligned with the time of the virus. This would be the first ceremony since the virus broke, with all three writers due to attend; there was excitement, too, because this year a new dimension would be added to the forest handover: a special room had been constructed in the heart of Oslo's Public Library, to keep safe the work set aside for a generation of readers yet to be born.

Since I'd first heard of Future Library, I'd longed to visit the forest grove where the books were growing. Each year, as a new writer was announced, I would wonder what message they would leave for the next century and what it felt like to assume that responsibility. Now, a few days before the ceremony was to take place, I was on my way – and running late – to meet

WILD CLOCKS

Ane Victoria Vollsnes, a plant biologist at Oslo University. We would travel by train to the Future Library grove, where Ane would tell me how the forest ecosystem is likely to change in the next hundred years. I wanted to find out how Future Library might also connect us to our own changing times.

I arrived fifteen minutes late, and was relieved to find Ane waiting patiently for me in the busy square outside the train station. Quiet and neat, she smiled politely at my bad and slightly breathless joke about our mismatched phenologies, and led the way towards the platform. Once we were on the train, I asked her about her research. Ane tracks how quickly Arctic ecosystems are changing; specifically, she studies the dual effect of air pollution and climate breakdown on plant species in Finnmark, in the very north of Norway. The Arctic is warming three times faster than anywhere else on Earth, she said. The chronoclasm is more fierce and abrupt there than anywhere else on the planet. Still, her face lit up when she spoke about the light and space of the far north.

Our train reached the end of the line, and we set off down one of the many branching paths through the forest. 'Most people have a kind of plant-blindness,' she said. All they see is green; but it's possible to learn a lot about how the forest tells time, if we pay attention.

Ane stopped and examined the bank on our left. Dozens of small-leaved bilberries were growing close to the ground, but scattered throughout, scorched-looking branches rose up clawlike, burnt orange against the green. Bilberrys have evolved to exploit snow cover, she explained, surviving the harsh Norwegian winter under an insulating blanket of snow. But in recent years the winter has been warmer than usual. Primed to expect snow to fall later in the year, the charred twigs had grown beyond the snow-line and perished.

I asked what other wild clocks we could find in the forest. We paused by a tall spruce a little further down the track. These trees coordinate their circadian rhythm with sensitivity to temperature, Ane told me, so as not to bud in December when it is too dark and cold. The new needles had been growing since May, and now each dark branch ended in a fat, feathery, light green tip. As the climate changes it might affect the timing of spruce phenology, or it may be that the spruce get new neighbours like beech, which extend their range northwards.

The forest around Oslo won't disappear, Ane reassured me. 'One hundred years is not a long time in the life of a forest. But even so, we could see a lot of changes in that time.'

Some plants that can exploit the increased levels of carbon dioxide in the atmosphere may grow more quickly, she went on. But the amount of nitrogen in the soil won't change, which means they may have more starch but not more protein – a potential problem for herbivores, especially in the breeding season. In Greenland, the caribou breeding season is precisely synchronised with the arrival of spring. Both calving and plant phenology are advancing as the climate warms, but at different speeds. Whereas calves are born, on average, just under four days earlier than thirty years ago, the plant-growing season has advanced by nearly five days. That single day's difference may not sound much, but for outliers within the range, the risk that their calves will starve is real, and the gap is widening, growing starker with each year. Reindeer in the north of Norway face similar pressures. During winter they rely mainly on mosses for sustenance, but in warm winters, when snowfall is punctuated by rain, ice can form impenetrable layers in the snowpack, sealing the mosses away from the reindeers' reach.

The Future Library grove stood off the main path on a steep, clear-cut hillside. Watched over by tall mature spruce, the young trees reached no higher than chest-height. Pink ribbons were tied to their upper branches and they wore the same bright tips of new growth, like hairy green caterpillars. In the years since planting, the scrubby slope had also been colonised by bilberries, raspberry bushes, and rowan. Ane pointed out the moose-nibbled top of a rowan tree, where it had poked out above the snow-line the previous winter.

We sat down on a rough wooden bench in the centre of the clearing, and shared coffee from Ane's flask while we listened to the hushed sound of hundreds of trees waiting to become a library.

How will this scene change, I asked, before the trees are cut and the books printed?

She paused before replying. One hundred years ago, there were cattle grazing in this forest, she said; maybe in the future wild boar would cross the border from Sweden and settle here – more new neighbours to accompany the northwards-marching beech. The balance between host trees and parasites might also be affected. White pine weevils spend much of their lives burrowed inside the trunks of living trees, feeding on the bark and laying their eggs. Their taste is for seedlings and young trees, and without protection up to half the newly planted trees could be chewed to pieces. Ane had helped devise a wax coating to protect the newly planted spruce from weevils. But in years to come, if the warming climate allows the weevils to breed multiple times each year, any trees seeded by the original grove may not be so fortunate. The children of the Future Library mother trees would then paint the grove in a rainbow of sickly hues, mixed among the healthy green. White resinous growths would spot their trunks where the insects burrow; the needles would turn red and drop until all that was left was a grey skeleton. Other colours would enter the forest: warming temperatures and shorter winters in British Columbia have allowed mountain pine beetles to mature faster and spread further. The beetles enjoy a symbiotic relationship with a blue fungus that converts nutrients in pine trees into food for beetle larvae; as they bore into pine trunks, the beetles draw the fungus further into the body of the tree, marbling the wood with a blue stain. If warming extends the range of the fungus-bearing pine beetles to the Nordmarka grove, the wood of fallen trees would be tattooed sky-blue.

The Future Library trees were safe, but the bugs can be patient. Weevils are especially attracted to the smell of cut spruce, Ane said. If, in 2114, the trees are cut during a weevil breeding season, the felling of the Future Library grove could precipitate a storm of pine weevils seeking places to breed. Printing the library could unleash a plague on the rest of the forest.

I had expected the library-in-waiting to feel like somewhere very obviously set aside. But it wasn't like that at all. Rather than timeless, it was very evidently full of time. One hundred years might not be long in the life of a forest, but Ane was right – so much change would take place in that span. The grove was immersed in the polyrhythmic flow of life, and as the tempos changed so would its sights and sounds (even, perhaps, smells – thinking about those wild boar). We were sitting inside a giant wild clock whose syncopated tick measured the evolution of the whole forest.

A few days later, I returned to the grove for the handover ceremony. Hundreds of people had travelled from the city to form a procession through the forest. Beneath the overcast sky there was a cheerful, holiday atmosphere. Children and pets scurried excitedly between the trees while the adults drank thick black coffee warmed on an open brazier. A microphone was set up by the bench Ane and I had sat on, now occupied by Katie and the writers who would be depositing their work. As an expectant hush descended, the wind in the tall trees whistled softly in answer.

'We're standing in this forest, an ordinary forest,' Katie said in welcome. 'But it's a forest filled with promise.' It became clear that the primacy of the clock held no sway here. In a message read on his behalf, Ocean Vuong (who had tested positive for Covid and was unable to travel) addressed the readers of 2114. 'As a Buddhist,' he said, 'I believe that my dying will propel me toward you, so that I may be among you when these texts are revealed.' 'One billion people believe time runs backwards,' remarked Tsitsi Dangarembga. Many African cultures think of time in terms of Zamani (the past) and Sasa (the present and immediate future). 'The present exists only to ensure that one moves back into the past well.'

'There is no past or future in art,' Picasso once said. But art has always reflected our changing relationship with time. Some of the earliest cave paintings – handprints in red ochre – record our ancestors' shift from the perpetual present of sensation to time as duration: from *I am here* to *I was here*. One of the gifts of this locked library is to extend this potential in art to act as a marker of presence in time. I once asked my friend, the Australian writer James Bradley, what message he would leave for future generations. 'We know you're there,' he replied. Future Library proclaims the same message: *while we were here, we knew you were there*. Just like the future it is promised to, the library is locked to us – we in our time cannot know what either holds – yet it links our time with those who are yet to be born, and by extension to the generations who will follow them.

'What tense would you choose to live in?' the poet Osip Mandelstam once asked his journal, before answering his own question: 'I want to live in the imperative of the future passive participle – in the "what ought to be".'

Future Library is fundamentally about how time is made together, I realised, in all the wonderful diversity of times we live by. And it includes the time of the forest itself. The books waiting a hundred years to be read connect us with people who have yet to be born, but they also invite us to synchronise our time with that of the trees waiting to become books.

The forester who was responsible for the grove stepped up to the microphone. He spoke with the shy pride of a new parent. 'Be careful of the trees!' he teased the crowd.

The spruce were now established, he said, their root systems healthy. Other species – rowan and birch – had begun to arrive. 'Now is the time to form the future forest – do we clear away these pioneers, or do we let them stay?' Clearing would create a dark stand crowded with only spruce; leaving the airier rowan and birch to flourish would break up the denser trees, allowing light to filter through the space as the trees grew. The decision would have implications for what the ceremony would look like in the future. He turned to Katie: 'What should we do?'

It is in the nature of rituals to grow and change, as the world they connect us with changes. In years to come, the handover – a human note struck once a year in the cadence of the forest – will adapt to the forest's changing shape. As it grows, so will the means we use to mark each new deposit in the library. The clear-cut slope we stood in formed a natural amphitheatre, but as the trees grow taller they will crowd the space, breaking lines of sight and limiting the number of people who can attend future occasions. Hundreds might shrink to a select few; or perhaps it will mean that the focus of the gathering shifts, from the authors to the trees themselves.

In the future, the trees will dictate what form the ceremony takes, determining 'what ought to be' for those who keep the ritual.

Before I left Oslo there was one more thing I wanted to see. This time, I made sure I wasn't late.

In February 2016 a pile of 200 decapitated reindeer heads appeared outside the district court in Finnmark. As snow fell on the raw and bloodied heap, inside the court a Sámi reindeer herder called Jovsset Ánte Sara initiated proceedings against the Norwegian state, which had decreed that his herd should be cut to no more than seventy-five reindeer. In 2013, the Ministry of Food and Agriculture had announced the implementation of tvangsreduksjon: the forced reduction of Sámi reindeer herds. For decades, southern politicians had worried that excessive grazing by reindeer would degrade pastures, resulting in the collapse of the delicate tundra ecosystem. Herders like Jovsset argued that such small herds would be too vulnerable to survive, placing the entire Sámi culture of reindrift, or reindeer husbandry, at risk.

The grisly pyramid was the work of Jovsset's sister, Sámi artist Máret Ánne Sara. She called it *Pile o'Sápmi*, after another pile of bones, an infamous photograph of a thirty-foot-tall mountain of bison skulls taken in Rougeville, Michigan, in 1892. The massacre of North American bison, a slaughter of almost incomprehensible fury, brought the animals to the brink of extinction, reducing a population of 60 million to less than 500 in just a few decades. Sara's artwork changed shape several times as her brother's case progressed, appearing like a spectre outside courtrooms across Norway, until she arrived at the final form: a curtain of 400 reindeer skulls, alternately weather-stained and bleached white, arranged to mimic the pattern of the Sámi flag. *Pile o'Sápmi Supreme* debuted outside the Norwegian parliament in December 2017, as Jovsset finally lost his case against the forced reduction of his herd.

I arrived early at Oslo's new National Museum, a vast, slate-grey modernist box situated in a sun-drenched courtyard, because I wanted to see *Pile o'Sápmi Supreme*, now hanging in the museum's foyer, for myself. The museum had opened just a few days ago, and visiting time slots were strictly allocated. As I approached I could see that a long queue had already formed outside, as people waited excitedly to see inside for the first time.

Slowly, the line made its way forward and I stepped gratefully from the glare of the courtyard into the shady interior. It took my eyes a moment to adjust, and then there it was: suspended on the far wall of the immense lobby, the curtain of skulls glowed eerily.

I wasn't prepared for the scale of it. It seemed truly epic; heavy with the weight of 400 reindeer, but also evanescent, ghostly. The bleached skulls' empty sockets seemed to stare back intently; every one of the 400 heads was perforated by a bullet hole like a third eye. Up close, they were surprisingly singular. Each animal would have been known to its herder, its personality and temperament a matter of simple, intimate knowledge, and even in death they remained distinct, as if the fact of being known had imprinted itself somehow on their bones. Together they flowed down the wall, nose to neck, ending in a ragged hem.

Minutes passed. For a long time, I couldn't take my eyes from it. I could almost feel the warmth and smell the stink of the herd, close-pressed on the open tundra. But, oddly, I couldn't hear it. Silence had wrapped around the skulls.

In W. H. Auden's poem 'The Fall of Rome', a modern empire crumbles. Clock-time begins to collapse: train carriages stand abandoned in desolate fields while whole cities are stilled by influenza. And yet, he writes, life and time persist:

> Altogether elsewhere, vast
> Herds of reindeer move across
> Miles and miles of golden moss,
> Silently and very fast.

Something in the skulls' silence spoke of the time that they had once inhabited, as if the same tremendous power they'd had in life, vast herds pressed nose to neck on the open tundra, flowed through them still. But their place of arrest, on the wall of a museum in the centre of Oslo, was also another kind of chronoclasm. The colonisation of time has not only deafened us to the rhythm of wild clocks, it has also eliminated – or at least sought the elimination of – many of the indigenous ways of timekeeping that would not coordinate with the priorities of empire-building.

Samantha Chisholm Hatfield, an enrolled member of the Confederated Tribes of Siletz Indians and a cultural anthropologist, writes of how indigenous concepts of time can vary widely, but are always deeply entwined with a sense of how time is made together with the places in which people live. Inuit Qaujimajatuqangit, the accumulated knowledge of the Inuit of Alaska and northern Canada, teaches that there are at least six seasons, related to changes in ice cover, hunting and migration. The Anishinaabe calendar follows thirteen moons in its seasonal round. Many indigenous cultures view time fundamentally differently to the future-oriented arrow of the Western imagination. Speakers of Aymara in South America and the Tuvaluan language group in the Pacific conceive of the future as located out of sight, behind a person. In others, the distinction between past and present collapses. In Ojibwe, 'aanikoobijigan' means both ancestor and descendant.

In the case of the Sámi, the Norwegian government's official time, with its inflexible concept of 'optimal herds', collided with the herders' concept of 'jahkodat', a northern Sámi term that conveys the seasonal distinctiveness of any given year. One year, so the saying goes, is not brother to the next. In spring, Sámi herd their reindeer from winter pastures on the inland plateau to the warm, green pastures along the coast. But the timing of this migration involves a complex calibration of factors, accounting for the fact that spring or winter may arrive sooner or later in any given year. Climate breakdown is exacerbating this fluctuation, and tvangsreduksjon would impose a limit on reindeer numbers that makes them vulnerable to the slightest seasonal variation. Nonetheless, jahkodat persists. Pinned to the wall of the museum – a place of preservation, where time is only permitted to stand still – the skulls spoke loudly of an understanding that time is made between people, animals and place.

The Potawatomi scholar Kyle Powys Whyte calls this 'kinship time', a way of understanding and experiencing time through relationships with other living things. When we think of the changing climate in terms of linear time, he writes, the image of a 'ticking climate clock' pushes us to respond in ways that cause harm – like imposing a cull on reindeer that also risks the Sámi way of life. But when time is felt as kinship, that anxious ticking goes away. Instead, we perceive duration, history, even the future, in terms of the health and strength of our relationships with other beings. This isn't to say that kinship time is any less concerned with the urgent need to address climate breakdown; rather, Whyte says, it 'reveals how today's climate change risks are caused by people not taking responsibility for one another's well-being'.

WILD CLOCKS

I'm not a Sámi reindeer herder; jahkodat – time made between herd, herder and the places they pass through – isn't my time to live by. What we in the West have lost in terms of the richness of telling time cannot be restored by repeating history and plundering other cultures. But perhaps I can discover a kind of kinship time by letting wild clocks adjust my sense of connection. Kinship time is time made together, even when the rhythms become disordered; something like a veery clock, coordinating my time in Scotland – where the effects of climate breakdown are modest – with the fiercer, faster changes bearing down on the Arctic summer might instil a sense of time that is kin to Potawatomi kinship and Sámi jahkodat.

By syncing with wild clocks like those in the Future Library forest, we might redesign the systems and infrastructure that sustain modern life. We are trapped in cycles of consumption that spin a fantasy of timelessness, of action without consequence, while 'forever chemicals' and non-biodegradable plastic pollute rivers, soils and groundwater. But forests have no concept of waste. A forest is a loop, in which dead matter is fed back into the system: past feeding future becoming past. Establishing a circular economy would not just involve a radical shift in how we manage materials; it would require a dramatic change in how we think about time. In forests, time is shared; we, on the other hand, tend to hoard it. The drive to pull as much as we can from the Earth, whether it is minerals, oil or nutrition, is a kind of temporal stockpiling.

One way to cultivate a sense of forest-time would be through soil. Soils are composed of time: the processes of weathering and decomposition that make soil are so slow that geologists speak of 'soil time', a distinct temporal category that operates well outside of human perception (as mentioned, it can take up to a thousand years to make a centimetre of soil). But soil is also composed of time in the billions or so microorganisms that can be found in just a single gram, each with their own time. There is much we could learn from microbial time: microbes live simultaneously on very long and very short timescales, being both extremely resilient and highly responsive to changes in their environment.

Our failure to appreciate soil time has grim consequences. Sixteen per cent of soils worldwide risk being exhausted by industrial agricultural practices within a hundred years, but well-managed soil can be farmed for up to 10,000 years. Soils are also carbon stores, but warming increases soil respiration, increasing the amount of carbon exhaled back into the atmosphere. Coordinating our time with the needs of soil would mean changing agricultural practices to sequester carbon and maintain soil health, for instance by rotating crops, avoiding pesticides, using seed drilling techniques that preclude the need for ploughing, and cultivating mycorrhizal networks that lock carbon in the ground; it would also mean cultivating a sense of time that balances resilience with receptiveness to change.

We might even redesign political time. Imagine what could be achieved with a political calendar that was set by the wobble in the jet stream or the faltering migration of butterflies, rather than election cycles. In *The Good Ancestor*, Roman Krznaric proposes four design principles for a political culture with the capacity to think long-term: political institutions charged with safeguarding the interests of future generations; citizen assemblies where long-term policy can be shaped; legal mechanisms to guarantee intergenerational rights; and the devolution of power from nations to cities, to obviate the worst excesses of short-term thinking. Reminiscent of Future Library, Japan's Future Design movement puts many of these principles into practice. Citizens of Japanese cities such as Tokyo and Kyoto gather in assemblies which divide into two groups: one represents the interests of the present; the other, the interests of 2060. The latter groups typically devise much more radical solutions, and in 2019 the city of Hamada decided to formally integrate the approach in city planning.

Redesigning our political system along these lines will be made significantly easier if we first learn to coordinate our time with the living world. Our cities, which might seem to be places where the clock rules most forcefully, are in fact where we can find the greatest diversity of wild clocks. Some are being smothered by human-made zeitgeber, like the urban songbirds serenading the false dawn of electric light. But those wild clocks are still there, expressing times that are wholly other to our own. Synchronising with these times is essential for making the imaginative leap that aligns the needs of now with those of the future. Properly integrated in the chronobiology of its local ecosystem, a devolved city-state with a citizen assembly of 'future guardians' could finally break out of the bubble of clock-time, and into a world of polytemporal relationships. But to guide and guard this transition, we need new ways of thinking and acting in time.

In practical terms, we might discover this in ritual. Krznaric notes that, in the Future Design movement, representatives of 2060 wear special ceremonial robes, 'to aid their imaginative leap forward in time'. To enter into ritual is to shift our tense, stepping into a time that is slightly removed from the current of everyday life. When wild clocks fall out of step they also shift tense, from the simple present of 'what is' to the imperative of the future passive participle, Mandelstam's 'what ought to be'. This new tense carries with it a kind of temporal ache, an acute sense of loss. But it is also the tense of hope and of ritual. Rituals both mourn what is lost and exhort what is to come. Rituals, writes the poet CAConrad, usher us into a moment 'where all of time is suddenly present'. A ceremony of What Ought To Be would bring past and future into alignment with our calamitous present, in a festival of remembrance and renewal.

Ceremonies of What Ought To Be could help us embed kinship time in our social structures and institutions. Each UN Climate Change Conference already includes a Young and Future Generations Day, intended to remind decision-makers of the obligation they bear to people who, yet to be conceived, will be born 'out of time' into a world where our presence and influence still shapes the climate. But local versions of this that are tied to the losses and kinships of a particular place, like Future Library, would give everyone a means to navigate the slippage in time we all inhabit, consciously or not; even if what is remembered is an animal's total eviction from time and place.

In the late eighteenth century, oysters were so abundant in the estuary where I live that three-man clinker-built boats would take 30 million each year. Two would row while one controlled the dredge, a net attached to an iron frame that dragged along the seafloor. To keep the dredge's mouth open, the boat had to maintain a constant speed: too fast and the dredge would skip and lift off the bottom; too slow and the mouth would close. Fishermen would keep an even pace by singing a 'dreg song', the asymmetric rhythm of which – five syllables sung by the leader followed by three repeated by the rowers – matched the greater effort needed to pull the oars through the water and the lesser effort involved in bringing them round through the air. This lopsided chant, they believed, would charm the oysters from their beds. 'The oysters are a gentle kin,' went one song. 'They winna tak unless ye sing.'

The dreg song's success was calamitous. By the mid-nineteenth century the annual catch had declined by more than 99 per cent. Oysters were declared extinct in the Firth of Forth in 1957, by which time the dreg songs had been silent for decades.

A ceremony of What Ought To Be on the shores of the Firth of Forth would recall not just the smothering rhythm of the dreg songs, but also the rhythms they stifled. As a keystone species, the oysters provided clean water and habitat for many other estuarine residents, a kind of bass pulse underpinning the river's various beats and tempos. Recent research has also suggested that oysters, like veerys, are climate clocks. Their sensitivity to changes in temperature and the flow of nutrients means they act as sentinels of climate variability. Yet, along the Forth, that warning system has been lost (although a small oyster population has tentatively begun to recolonise the estuary, it pales next to the vast beds, covering 100 square miles, which once blanketed the estuary floor). All phenological mismatches are steps towards extinction, straying from a carefully kept rhythm, and risk unravelling entirely if pushed too hard. The river where I live has slipped into the same disjointed time that threatens Australian mountain pygmy possums, Greenland caribou and a forest outside Oslo.

'This is a library built by many hands,' Katie Paterson said during the Future Library forest ceremony. By safeguarding knowledge against loss, libraries coordinate our time with that of future generations – all libraries are statements of faith that, in the future, people will still value the insights of science and art. With a forest-clock, we can synchronise with an entire ecosystem, calibrating our tempo with nature's intricate groove. Our days should be told by forest-time and oyster-time; by the hastening of the Arctic spring and the loosening of the bonds by which species make time together.

At the museum in Oslo, as the minutes ticked by, I began to feel the pull of the clock on my attention once more. It was time to leave. With one final glance back at the gleaming curtain of reindeer skulls, I picked up my bag and walked towards the exit.

On the day I arrived in Norway, I had gone with Katie to visit the room where the unread works would be stored. We were greeted at the train station by Anne Beate Hovind, the Future Library producer, who beamed in welcome and – being several inches taller than either of us – enveloped Katie and then me in a warm hug. Just a few days earlier she had signed a contract with the city of Oslo, formalising a partnership for the city to steward Future Library for the next ninety-two years. Tears stood in her eyes as she spoke about it. 'I never expected politicians to take such a long view!'

The Silent Room was located on the top floor of Oslo's busy public library. At Anne Beate's request, we removed our shoes before we stepped into the dark interior.

A short and narrow passage curved gently into a small rounded room, only just large enough for the three of us. The whole room, walls and ceiling, was made of strips of pine arranged in horizontal layers. The air smelled rich and woody. It was a bit like entering a giant beehive, or even the inside of a tree, except for the shards of what looked like ice studded throughout the room and passageway. Irregularly arranged floor to ceiling were the 100 hand-made glass drawers that would keep the Future Library works safe. Lit from behind, they filled the whole room with an ethereal glow.

I once visited a Neolithic tomb in Orkney. I'd had to crawl through a narrow passage that opened out into a chamber very like this one, although dark rather than light, the walls layered in stone not wood. Passing through it was like being born in reverse, emerging in a womb-like space that seemed to be set aside from the regular flow of time. Similarly released from the rule of the clock, time in the Silent Room felt fuller and more open, and not entirely ours; not something we control but something that is made (and told) as we live and work together. Meanwhile, deep in the quiet forest beyond the city, the wild clocks in the Future Library grove were keeping their own time.

I asked Anne Beate how it felt to depend on a generation of library stewards who have yet to be born. She grinned broadly. 'We have a saying in Norwegian,' she said. 'I think you have the same in English. Ta det som det kommer, "take it as it comes". That is right, yes?'

She sat back, smiling. 'Ta det som det kommer,' she said again.

'Take it as it comes,' I replied.

The Lion-Man of Hohlenstein-Stadel

7

THE LION-MAN'S LEAP

How synthetic biology could save vulnerable species from extinction

One day, around 40,000 years ago, someone picked up a piece of mammoth ivory and began carving it into a creature that had never existed before. The lower portion had the body and limbs of a human, but the head and forearms had a cave lion's grace and power. At some point lost to history, the totem fell out of use and into a long darkness, until August 1939, just days before the start of the Second World War, when it was discovered in fragments at the back of a cave in what was by then southern Germany. In the following decades, subsequent excavations unearthed more fragments, and for the first time since before the last ice age, the Lion-Man of Hohlenstein-Stadel stood upright again.

He stands just over thirty-one centimetres in height. Mottled with age, his ivory skin still has a warm glow, but the decay of collagen has broken the material along natural growth layers, leaving him looking like a partly peeled tree limb. There has been some debate over the Lion-Man's gender: male cave lions don't grow manes, and the Lion-Man's triangular genital plate is ambivalent. But despite some archaeologists arguing that the Lion-Man is a Lion-Woman, most research now refers to 'him'. He is what archaeologists call a therianthrope: hybrid animal-human figures that emerged around this time in the Paleolithic imagination.

Animals would have dominated our ancestors' world, so attuned to the presence of predator or prey that familiar sounds and shadows haunted their dreams. It is easy to suppose that living so close to the lion's strength or the bird's swiftness would translate into a dream of adopting those traits themselves. Perhaps this is the inspiration that lies behind the making of the Lion-Man. A mix of desire and fine-grained attention is visible too in the slightly tilted angle of his head, the snout pointing left while his ears turn rightwards, as if his attention is snagged on a particular sound. Behind the right ear there is even a furrow, where small muscles contract in concentration. His shoulder blades are drawn together and he seems to stand up on his toes, as if poised to leap. But into what?

The Lion-Man proves we have been imagining impossible creatures for a very long time: in fact, long before domestication began, our ancestors practised a kind of imagined synthetic biology. We too are poised to leap into a new chimerical age, where nature is no longer limited to evolution's random shuffle, and traits and attributes can be cut from one organism and pasted into another. Gene-editing technology has existed since the 1970s, and in that time synthetic biology has brought remarkable advances in medicine and agriculture; it could yet represent a vital tool in the struggle to save species under threat of extinction.

Most of the changes driven by human action are unintentional. Our cities are evolution engines by accident, not design; we did not flood our oceans with noise or the atmosphere with carbon because we wished to rewrite whale songs or detune wild clocks. But the history of our accidental interventions in evolution is bookended by what biologist Beth Shapiro calls 'the power of human intent as an evolutionary force'. Domestication, and the distinct phenotype it gave rise to, had as great an effect on what it means to be human as it did on what it means to be a dog or a cow or a horse. Now, in the age of synthetic biology, the power of human intent could perhaps undo much of the damage we have done since.

For Shapiro, a clear line runs from the laboratory back to the first farms. But the problem with selective breeding, she told me via a video call to her home in California, is simply that it's messy. 'Half the DNA gets thrown away on the evolutionary cutting room floor. Synthetic biology is doing the same thing we've been doing for generations, but in a much more deliberate and precise way.'

Precision gene editing became possible with the invention of a technology called CRISPR/Cas 9. Clustered regularly interspaced short palindromic repeats (CRISPR) are DNA sequences that evolved to identify bacteriophages, viruses that prey on single-celled organisms. In 2012, Jennifer Doudna and Emmanuelle Charpentier discovered that it was possible to carry new DNA into a genome via these repeating sequences, using a CRISPR-associated protein called CAS-9 and an RNA guide molecule. CRISPR is often likened to molecular scissors, so fine that even single letters of DNA can be exchanged. The RNA molecule ensures the hand wielding the scissors never slips, and the protein makes the cut. CRISPR guide sequences can be designed to make an edit at any point in the genome, and thus individual genes switched on or off.

However, in most cases, the leap from genotype to phenotype is beyond us, Beth said, because we don't yet know which gene or combination of genes to edit to effect a particular change. But where the knowledge exists, the potential is vast.

The American chestnut once dominated the landscape of eastern North America, but a fungus introduced at the start of the twentieth century by shipments of ornamental Japanese chestnuts proved so virulent it killed billions of trees. By the middle of the century, the American chestnut was functionally extinct, restricted to a small number of remnant populations in the Pacific Northwest and a still-intact root system which the pathogen would attack whenever shoots appeared above ground. Selective breeding has produced a hybrid, borrowing immunity from a Chinese chestnut, at the cost of 15 per cent of the original genome. But by editing just a single gene in the American chestnut genome, scientists have broken the lethal link between pathogen and tree. The gene in question comes, not from another tree, but from wheat: the wheat gene manufactures an enzyme that suppresses the fungus, allowing tree and fungus to coexist. Since 2019, transgenic American chestnuts have been planted at sites across their original range, reviving a lost ecosystem and restoring a vital food source for woodland species like black bears and white-tailed deer.

With precision gene editing we could redress the ecological imbalances caused by introducing non-native species to new habitats. It may help us to rescue species at risk of extinction, or reset the wild clocks of species whose sense of time has been unsettled by climate change. Microbes could be programmed to sense pollution and consume toxins. We could safeguard nutrition against climate breakdown, improving photosynthesis in crops and introducing heat tolerance to the DNA of cattle, or even forsake our dependence on dairy altogether in favour of proteins derived from gene-edited bacteria. Gene-editing technologies are extraordinarily powerful tools. We can use them to crack open the code of life, to tinker with its message, or splice together genetic information separated by millions of years of evolution. The prospects are dizzying; but while these technologies have profound implications for what the nature of the future may look like, just as profound are the ways we may be changed by using them. A tool always alters the one who wields it. From the stone knife to the internet, each new invention has shifted how we see the balance of possibility and impossibility in the world.

When I asked Beth about this, she cited Stewart Brand's famous maxim: 'We are as gods and might as well get good at it.'

'We have created an environment in which a lot of species just cannot survive,' she said. 'And it's our fault. If we want a world that is both biodiverse and filled with people, then we have to get better at this.'

What does it mean to get 'better' at playing god? Gods can be beneficent, but mythology teaches that they can also be capricious, determined to impose their will and disinclined to either remedy their mistakes or learn from them. With synthetic biology we could undo so much damage, but if our idea of ourselves as godlike remains unchallenged and unchanged, we risk repeating the same mistakes. The existence of patents on the bodies of genetically modified broiler chickens, turning the animal's very being into intellectual property and its genome into the preserve of lawyers and investors, suggests that our old habit of seeing ourselves as masters of creation will die hard.

To avoid this fate, we could do worse than to look to the Lion-Man, where we find our ancestors beginning to imagine that they could mould life into a new shape. The Lion-Man may represent a Palaeolithic god – or he may be a man in disguise, a shaman dressed in the head of a lion. The people who made him may have seen god, man or beast in the figure they carved. Most likely, they saw a mingling of all three. Gods tend to live apart, separate from the rest of creation, but the Lion-Man represents a union, an affinity between human and animal; it is out of this fellowship that something greater, a sense of the divine, emerges. Whatever changes we make to other species, we should be prepared to look again at ourselves.

Where gods often seek to control, nature tends to work together. It's thought that it would have taken around 400 hours to lift the Lion-Man out of the ivory tusk. Most archaeological research refers to a single carver, but I find it more convincing to imagine he was the work of a community, his form taking shape as it passed through many hands, each person holding in their mind the impossible image and feeling its weird contours alter their notion of what could be. The Lion-Man's leap was taken together.

The Lion-Man's first lesson is the most elemental: ours is a chimerical world, part wild, part made, much maimed; and almost all of life emerges out of some form of collaboration.

Perhaps the best example of chimerical living in nature is found in the oceans. In appearance, corals seem to straddle the boundary between plant and mineral. Healthy reef gardens are alive with tropical hues, but their stone fronds and leaves are sharp enough to strip away skin. Coral biology is equally chimerical, an ancient coalition between animal and microbe. Polyps, the tiny, soft-bodied animals whose calcium carbonate secretions build the vast underwater conurbations we know as coral, rely on zooxanthellae – microscopic dinoflagellates or algae – for nutrients and energy. Zooxanthellae also produce the photosynthetic pigments which paint coral cities in the colours of an acid dream. But this alliance is also extremely delicate –

vulnerable in particular to changes in temperature – and is being steadily picked apart as our broken climate spills heat into the oceans. Heat brings on a kind of nausea in the polyps: the zooxanthellae feed the hosts a surfeit of oxygen, causing them to vomit up the dinoflagellates, leaving the coral stripped of colour and starved of food. And without coral, a whole host of marine animals that depend on it for shelter or food – up to a quarter of all life in the oceans – are also at risk.

We are imposing a fatal separation on tropical corals around the world. Bleaching events, where a spike in water temperature kills off whole swathes of reef, have doubled in frequency since the 1980s, and are projected to be a near-annual occurrence by 2075. What is more, they are getting more severe with each wave of heat, like a fever that returns with renewed ferocity each time the patient relapses. The heat stress that led to the 2024 coral bleaching event was the highest ever recorded. The rapid warming of the oceans is exerting an extreme form of selective pressure on tropical corals worldwide. A few may be able to evolve sufficient heat tolerance to keep pace, but for most the future looks bleached of hope. The gulf between the speed of change in the environment and the speed of change in the organism is too wide; but with assistance, it is possible that coral might just make the leap.

In 2012, ecological geneticist Madeleine van Oppen and marine biologist Ruth Gates realised that, if coral was to survive, we had to intervene in its evolution. They devised a form of assisted evolution: a suite of techniques, including selective breeding, assisted gene flow, and gene editing, which could force the pace of coral evolution to keep up with the rising heat.

'I don't want to control nature,' Madeleine told me over a video call. 'I just want to help it get over that hurdle.'

The idea was to step into the breach in the coalition between polyps and their dinoflagellates, temporarily suturing the relationship – at least until we can act to bring down emissions and turn down the heat in the oceans. As farmers improve the size of their tomatoes by selective breeding, it is possible to selectively breed corals for heat tolerance, Madeleine explained. Even during the worst bleaching events, some species show a degree of resilience, and – because heat tolerance has a genetic basis – cultivating this could spread tolerance throughout a population. Corals that have already adapted can be transplanted to vulnerable reefs, boosting the level of heat tolerance in the gene pool. Or corals might be helped to adjust to the coming climate. 'We call this conditioning or hardening,' she said, 'where you expose corals to stress that doesn't kill them.' Like a boot camp for corals, the process switches on a latent ability to withstand high temperatures. The conditioning is done in a lab, before moving the heat-hardened corals to the wild. The same methods could be applied to the coral symbionts, exposing the microalgae to stress or crossing different strains from different locations to engineer an extreme phenotype. But results varied for each method as they tested them, Madeleine told me. Hardening was a precise process, probably involving switching on or off multiple genes, which they hadn't yet worked out how to do consistently.

Assisting evolution may sound like exercising a divine power, Michelangelo's God conjuring life with a single gesture of the hand, but in reality it is difficult, patient work. Madeleine carries out much of her research at the National Sea Simulator, an artificial coral city in a laboratory in Queensland, Australia, where they monitor and intervene in the coral 'spawnathon'. Coral spawning is triggered by the appearance of a full moon in a particular season (in the case of the Great Barrier Reef, October and November). Millions of sperm and egg bundles are released into the water in vast billows, rising to the surface like an orgy in a snow globe; when the larvae fertilise, they sink back to the ocean floor, settling new coral colonies. Working within the narrow window set by the corals' lunar zeitgeber, dozens of researchers move corals by hand from tanks to buckets to large spawning trays, then wait, glass pipettes and assay plates in hand, for the carnival of coral sex to begin.

Madeleine told me they were experimenting with cultured coral larvae and symbionts in the lab, 'playing around with them' to create mutations that may yield greater heat tolerance and pairing different lineages to see if an extreme phenotype emerges. It's a process of trial and error: the technology for assisting evolution is new, and there are many gaps in knowledge. Heat resistance may involve not one or two, but perhaps hundreds of genes; and the route from altered genotype to a specific heat-resistant phenotype has yet to be mapped. Traits selected for heat resistance may not be stable once introduced to the wild, or may require costly trade-offs (for instance, increasing heat tolerance may impact how well the coral can process nutrients). And then there is the problem of scale.

The Great Barrier Reef covers 214,000 square miles and comprises billions of individual animals and around 600 different coral species. Selective breeding or gene editing of the coral host, the polyp, would ideally happen at the fertilised egg stage, so that any mutation would spread through the colony. 'Now think about it, how do you upscale that to effect all the offspring you want to put on the reef?' Madeleine said, with a rueful laugh.

Most coral scientists seem to agree that assisted evolution would be most effectively practised by communities living closest to threatened coral reefs. After all, people who live alongside the coral have the best opportunity and the greatest motivation to help it survive; and on the Great Barrier Reef, those who have lived alongside it for the longest are leading the way.

For upwards of 60,000 years, the Yirrganydji have lived in Dawul Wuru, an area of around 200 square miles of coastline in northern Queensland between Port Douglas and Cairns that was shaped by the passing body of Gudju-Gudju, the Rainbow Serpent, as it moved over the early Earth. It comprises both the rainforests and creeks of Pulmpa (Land Country), and the reefs and cays of Kul-Bul (Sea Country), which stretches for tens of kilometres from the present coastline to the continental shelf. Many reefs are sites of profound cultural significance, but Dawul Wuru borders some of the worst affected areas of the Great Barrier Reef. Something had to be done to protect this quickly vanishing heritage, so in 2010, the people of Dawul Wuru formed the Yirrganydji Land and Sea Rangers. The Kul-Bul project is a pilot reef restoration initiative, led by indigenous people in collaboration with marine scientists and the tourism industry. In 2022, Yirrganydji rangers managed the transfer of coral larva from Gunggandji Sea Country, a less severely bleached region further south, to Dawul Wuru. The plan is to deliver 30 million Gunggandji coral larvae to Yirrganydji Sea Country by 2025.

The Yirrganydji are one of over seventy Traditional Owners of the Great Barrier Reef. Many others have their own Land and Sea Ranger programmes, including the Gunggandji. These local coalitions, which bring together traditional ecological knowledge and Western science and are attuned to the precise make-up and conditions of their neighbouring reef, could help vulnerable coral to clear the hurdle of a rapidly changing climate.

'The Great Barrier Reef is a cultural landscape,' writes Gavin Singleton, coordinator of the Yirrganydji rangers, and it is one

THE LION-MAN'S LEAP

laced with creation stories that overlap culturally and geographically. Many of these stories carry messages about how to live in these environments, he told me over a video call. 'We tell a story of how an ancestor had been living out on the flat land, where the coastline was the continental shelf. Where the reef is now, then it was all forests and hills, and the rivers and creeks would run all the way out to the continental shelf. This ancestor did something that he shouldn't have done out there: he killed a fish that he wasn't supposed to kill, and this caused a disaster. The waters rose up and tried to drown him. So, he escaped from that area and fled to the highest mountains, and as he ran for his life he tossed the fruit and vegetables he was carrying to lighten the load. When he reached the highest point, he heated all these big rocks in a fire and cast them into the water. The water turned to steam which stopped it rising any further, and where the rocks landed they became the inshore coral reef system.' Some of the forests that were overtaken by the flood also became different reefs, Gavin explained, as did the vegetables that the ancestor threw aside as he fled.

The northern border of Darwul Wuru joins Mandingalbay Yidinji country, and one Yidinji Dreaming (or creation) story tells a similar tale of how the reef was made. To ensure harmony between everything that lived in Sea Country, Bhiral, the creator of the Earth, made a special fish that would maintain this balance, and prohibited anyone from catching it. But two brothers ignored Bhiral's injunction, spearing the fish and angering the god, who rained lava down on the people. In desperation, they picked up the flaming rocks and cast them into the sea, where they cooled and formed an immense barrier. The story of Bhiral emphasises harmony as the basis of a flourishing ecosystem, and the destruction that can follow when appetites run out of control.

Dreamings and the people who live by them are symbiotic: like the coral dinoflagellates that feed the polyps, they nourish the Yidinji connection to Mandingalbay Yidinji country and the Yirrganydji connection to Dawul Wuru. Biologically and culturally, the reef is composed of small but vital coalitions, which have sustained it through successive ages of the Earth. But these alliances must also maintain an essential balance: too much heat, and the corals sicken; neglect the contributions of Traditional Owners, and any effort to preserve coral will also preserve the domineering attitudes that made the crisis in the first place. As Red River Métis scholar Zoe Todd puts it, we can't simply 'add a dash of native knowledge to current approaches and call it reciprocity'. The individualism and false objectivity of Western culture must make way for the irreducible interconnectedness that is the foundation of indigenous world views. The Kul-Bul project has the Yirrganydji concept of Yurrbin Mamingal – or 'caring for our coral reef' – at its centre, and Madeleine told me that much of her work entails close and careful collaboration with Traditional Owners. It can be complicated, she said, but it is a vital part of their work. Permission to carry out research on any part of the reef must include the permission of the Traditional Owners; where boundaries between Sea Countries are involved, each group must be consulted.

Assisting evolution is, in the end, about forming alliances with other species: offering ourselves as a third party to the ancient collaboration between polyp and microbe, which has also thrived for millennia by embracing the lesson that together is all, and alone means dead. We hope these partnerships need only be temporary, lasting just long enough until corals have adapted to the coming climate. But the habit of alliance-building, whether it is between Western and indigenous cultures or between humans and other species, should not be temporary. Wherever we need to intervene in evolution, we should do so as partners in a process rather than autocrats; perhaps then there is a greater chance that we will also find ourselves changed in the process.

For some, assisting evolution is either too slow or too limited in scope; only the power of a god will do. But perhaps we should be careful what we wish for. What if we did have that power – a single instrument potent enough to direct the fate of an entire species? How to reckon with that burden of responsibility?

One day in 2013, biologist Kevin Esvelt realised it was possible to create CRISPR edits that would pass from one generation to the next. Every sexually reproducing organism inherits half its alleles (or DNA sequences) from each parent; every allele therefore has an equal chance of surviving into the next generation. But there are exceptions: gene drives are elements in the genome whose capacity to propagate is enhanced. They represent a strain of natural bias in the system, favouring certain genes for inheritance over others. Kevin realised that he could build a synthetic gene drive by adding CRISPR gene-editing machinery to the new material he cut and pasted into a genome. When the edit was introduced to one allele in an organism's reproductive cells, the CRISPR components would impose the same edit on the corresponding allele, guaranteeing that it would spread throughout the population. With a synthetic gene drive, the genome effectively edits itself according to our instructions: we could boost heat tolerance throughout a population, break the link between pathogens and hosts, or impose sterility on introduced species, driving them to extinction. It is one of the most momentous technologies ever devised.

Like something out of myth, Kevin realised he had discovered an instrument of awesome potential. 'The next day, I woke up in a cold sweat,' he said when I spoke to him about his invention. A gene drive could achieve miracles, like eradicating the plasmodium parasite responsible for malaria; but it could also become a terrible weapon. What if a rogue state wished not to cure but to impose malaria on its enemy, or sterilise their harvests? An unregulated gene drive could become a form of biological warfare.

Even without contemplating the more apocalyptic scenarios, a gene drive might make the ecological problems it was intended to solve infinitely worse. A standard synthetic gene drive could, in theory, propagate indefinitely; just as cane toads were originally meant to remedy the problem of crop-eating beetles, a a gene drive intended to do good could turn out to be as uncontrollable, even devastating, as the most virulent introduced species. A gene drive could be an extinction machine, giving whoever wielded it the power of life or oblivion over entire species.

Initially, Kevin kept his idea to himself while he came up with ways to use it safely. He devised a series of breaks or 'kill switches' so that any gene drive architecture would contain instructions to halt its spread. Using a 'daisy-chain drive', the edits could be limited to a fixed number of generations by splitting parts of the gene drive architecture across the genome, creating multiple break points. Each element would require the others to connect in order to drive the change, and as the connections broke down – an organism may inherit the molecular scissors but not the guide molecule, for instance – what was initially a flood of edited genomes spreading through the population would slow to an ebb. Daisy-chain drives dilute the terrifying potency of self-replicating gene drives, making it

THE LION-MAN'S LEAP

possible to create localised versions that essentially run themselves into the ground before they can spread out of control.

Daisy-chain drives might limit the awesome potential of the gene drive, but Kevin realised he had to do more than amend its architecture; he also had to think about the environments where it might be introduced, and the people who would have to live with the consequences. The only way to ensure the technology was truly benign was to start listening to people.

On the islands of Nantucket and Martha's Vineyard, up to 40 per cent of households are affected by Lyme disease, a bacterial pathogen carried by blacklegged ticks which, if untreated, can lead to heart complications and chronic pain. The ticks are in turn mostly carried by deer, but can also infect white-footed mice, which act like a pathogen reservoir, incubating the bacteria and increasing the chances it will pass to deer, and from deer to people. A gene drive designed to immunise the mice against the pathogen would drain the reservoir; the benefits of such a project were clear when it was proposed in 2016, but local people were anxious about the implications of unleashing such a powerful technology.

'I think it is a moral imperative that we give people a say in decisions intended to affect them,' Kevin said. Despite acknowledging the risks, the people of Nantucket and Martha's Vineyard said yes to the Mice Against Ticks project, which helped to develop a model of community-guided science that has since been used to address similar pest-related problems. In Burkina Faso, villages have consulted with scientists regarding the use of gene drive technology to suppress or even eliminate malaria in their locality. And in Aotearoa New Zealand, a synthetic gene drive could restore ecosystems that have been devastated by introduced predators.

Once, there were no rodents or mustelids in Aotearoa. Native birds nested on the ground without fearing for their eggs. The first people to arrive, from eastern Polynesia, brought kiore, a type of rat that drove snipe-rails and owlet-nightjars to extinction. But the real problems began when rats, mice and rabbits landed from European ships. With no evolved defences, local frogs, bats, birds and lizards began to slip beneath the rising rodent wave; what had been a gentle decline with the kiore became precipitous. As so often happens, the introduction of one non-native species led to the arrival of others to control it: ferrets and weasels were introduced to control the rabbit population, but rapidly became pests in their own right. Following 85 million years of biological isolation, Aotearoa now has one of the worst extinction records on Earth.

In 2016, the government of Aotearoa New Zealand announced a plan to be free from pests by the middle of the century. Conventional pest management usually involves traps, which can be slow to affect a whole population, or toxins, which can bring their own risks. But a targeted gene drive that used what Kevin calls the 'last litter' approach – where males are edited so that they cannot produce fertile daughters – could suppress the rat or weasel populations within a matter of generations, without the need for cages or chemicals. As he had found elsewhere, people were intrigued but wary.

In New Zealand, Kevin spent a lot of time listening to Māori. The main thing he learned, he said, was that gene drive can't be reduced to a simple question of designing the most effective technology. Make something that can work equally well anywhere and you may make a more powerful tool, but you can also lose sight of the place you're trying to change. Treating gene drive as a purely technical problem strips away vital insights that can only be found in a deep, abiding sense of belonging. Kevin told me he recognises that there are aspects of his invention that he, as tauiwi, or someone who is not Māori, could never fully understand. 'It made it very easy to approach the Māori and say, "We know that it takes decades to understand this. We're just going to have to trust you on it."'

Tame Malcolm is a Māori conservationist. He learned indigenous approaches to pest management as a child, and started work as a ranger as soon as he left school, trapping non-native species like kiore. He explained to me that the Māori world view rests on Mātauranga, an intricate system of knowledge and values which has evolved over hundreds of years and is contoured to the ecology of Aotearoa. 'I grew up where it was all one and the same. Science is Mātauranga and Mātauranga is science.'

When I spoke to Tame it was 5 a.m. in New Zealand. Having young kids in the house meant this was his quiet time, he said, smiling broadly. I admired his good humour at such a punishing hour. 'Oh, those are big questions,' he exclaimed at one point. 'This is good thinking for the morning!' A Māori-centred ecology, he explained, understands that humans are part of the ecosystem, rather than apart from it. A tremendous range of values and principles inform this, but two in particular are essential. The first is whakawhanaungatanga. Like many Māori concepts, it doesn't translate directly into English, but is perhaps best understood as a deeply ingrained process of forming and maintaining good relationships. 'That's Māori philosophy, that everything's related,' Tame told me. Whakawhanaungatanga connects individuals to family and tribe in bonds of cooperation and trust, and places human life in the midst of the great tangle of relationships that make up the living world. 'When my tribe came over from Tahiti to New Zealand, everyone on the boat needed to be for the family, to have that holistic mindset,' said Tame. 'So it's stitched into our DNA.'

The second concept is whakapapa. Whakapapa is the fulcrum around which Māori social life turns, and where whakawhanaungatanga is expressed most fully. Again, English doesn't have an equivalent, but Tame said genealogy is probably the easiest way to describe it, but in a much fuller sense than the Western notion of tracing a family line. Whakapapa describes the Ira, or inherited life, that every Māori carries within them. Ira is genetic, but it also has a spiritual dimension, as all Māori are descended from atua – the gods Ranginui (Sky Father) and Papatūānuku (Earth Mother).

'But whakapapa doesn't just have people in it,' Marcus Shadbolt told me. 'It also has the entirety of the natural world.' Like Tame, Marcus is Māori and works in pest management. His organisation, Te Tira Whakamātaki (meaning 'the watchful ones'), promotes indigenous approaches to conservation. He elaborated on the richness and depth of whakapapa. 'If you trace your whakapapa back far enough, eventually you get to birds and trees and everything else that makes up the natural world.' Whakapapa includes the environment that nurtures a person, from fish to forests; every Māori inherits it from their mother and father, and from the land into which they are born.

'And it doesn't stop at living things,' Marcus continued. Mountains and rivers form part of whakapapa – Lake Ōkataina, where Tame grew up, 'is part of my DNA', he told me. 'The lake is my ancestor.' And some take it even further: when introducing themselves, some elders will state their full whakapapa beginning with Te Kore, the nothing: the absence before creation. Rather than genealogy, Marcus prefers to think of whakapapa as a kind of taxonomy. 'It's a way of sorting the natural world,' he explained. The difference from Western taxonomy is that, instead of adhering to species boundaries, whakapapa is rooted in relationship. 'Instead of putting two birds together because they are closely genetically related, in whakapapa you would put two birds together because their niches overlap.'

The questions Māori ask about gene editing revolve around whether it will enhance or diminish their essential values. Improving the resilience of a species or an ecosystem with synthetic biology could enhance kaitiakitanga, the principle of guardianship that makes Māori responsible for the rest of the living world, but the possibility that a gene drive could run out of control, despite Kevin's carefully engineered breaks in the system, risks undermining this, making Māori irresponsible stewards. Some Māori might be sceptical of transgenic organisms like the American chestnut with its single wheat gene, because introducing a foreign gene could diminish its whakapapa; but simply editing a species' genome could enhance whakapapa if it contributed to its long-term well-being. But what counts as 'foreign' means something very different than in a Western taxonomy grounded in the concept of distinct species.

Once, Kauri, a coniferous tree that can grow up to fifty metres tall, and Tohorā, the southern right whale, were siblings who lived together on land. Tohorā longed for the oceans, however. When he saw the beauty of the deep water, he tried to convince his beloved Kauri to join him, but Kauri loved the soil and the sky, and would not be persuaded. Instead, they exchanged gifts: Tohorā dressed Kauri in his scaly skin which became rough bark, strong enough that his brother could grow taller than the other trees, and leaving his own skin so smooth he could glide easily between the waves. Kauri gave Tohorā oil to keep him warm and protect him from salt, and he taught the whale to sing.

'Kauri said, "I'll stay here and look after this place; you go off and explore, and then come back and tell me what you find,'" Marcus said. 'Our stories tell us that Kauri and the whale have the closest relationship to one another.'

In the 1970s, the kauri trees began to suffer from a fungal infection that attacked their root system. The pathogen damages the delicate tissues that draw nutrients into the rest of the tree, effectively starving it. Infection is usually fatal; by the turn of the millennium, it was killing thousands of trees each year, leaving their bleached skeletons poking above the forest canopy. 'When kauri dieback started attacking the kauri tree, the first response of Māori was to go see what the whales were doing,' Marcus said. Some said that grieving whales were drawn to help their stricken brothers, which explained why so many were beaching themselves. Then Māori healers, or Tohunga, discovered that a balm made from the bones of beached whales could heal the infected trees.

Whalebone appeared to be the first successful treatment for kauri dieback, Marcus said. 'So if you told someone that you could save the whales by using genes from a kauri tree, you'd have no opposition from Māori, because to them you've only gone back one generational step.'

By supplanting species boundaries with the bonds of ecological relationships, Whakapapa speaks to the possibilities of the new chimerical age. However, there's no single Māori point of view on gene editing, Tame said. Every iwi, or tribe, might see the issues differently; disagreements could crop up between generations or within a single family. This was Kevin's experience as well. The two iwis that he engaged with most were Ngāpuhi on the densely populated North Island and Ngāi Tahu on the more sparsely populated South Island. Their attitudes towards gene drive differed considerably: whereas Ngāpuhi generally preferred to trust in traditional approaches to managing pests, Ngāi Tahu were more open to gene drive. Both wanted to ensure that future generations would maintain their tie to the land, but whereas on the North Island that meant maintaining traditions and professions that rely on close human supervision, many Ngāi Tahu, with fewer people and much more land, felt it wasn't practical to remove non-native rodents with conventional tools.

What unifies Māori is an understanding that synthetic biology can't just be seen as a technological challenge, or even judged from a drily ethical perspective. Whakapapa and whakawhanaungatanga touch on every part of what it means to be a living being in the world who is connected to other living beings. Whenever a new idea or invention is proposed, 'if Māori can't see that family or holistic approach, it will probably get cut down real quick,' Tame explained. 'And before you really dive into something, you have to understand its Whakapapa.'

One way to think through the ethics of synthetic biology, according to environmental writer Emma Marris, is to ask whether the animals themselves would vote for their genomes to be edited. The absence of an interspecies translation engine makes this impossible, but asking after a creature's Whakapapa could be the next best thing. All living things have whakapapa; in the case of other animals, it is the evolved relationship between body, behaviour and place that fits the animal to its environment and the other creatures it lives with or depends upon. To establish an animal's Whakapapa involves tracing its relationship back to the atua that rule the place where it lives. Every place is ruled by atua, who establish kawa, the principles that determine what counts as right behaviour in that place. Kawa set by the god of the forest show how to keep the forest healthy; kawa set by the god of the waters guard the well-being of the rivers and seas. Knowing an animal's atua and whakapapa means you can work out what can and can't be done with it, Tame said.

This included non-native species like rats and possums. Māori distinguish between taonga, or treasured species – those that were in Aotearoa when the Polynesians arrived – and the species that arrived with European settlers. 'Now, some people might say, possums aren't from here, so they don't have whakapapa,' Tame said. But Māori have always understood the need to be flexible in how they relate to the world around them; that what was does not set a limit on what could be. When the first Māori arrived in Aoteoroa, they named a new god, Tāwhirimātea, the god of weather, to account for the violent, unpredictable winds of Aotearoa, so wildly different to the winds they had learned to navigate by on their home islands in the Pacific. Introduced species like possums are no different, Tame explained. After generations of living in Aoteoroa, they have acquired whakapapa that can be accounted for.

Moreover, these invasive species, like the possum, are taonga to the indigenous people of Australia, he said. Just as we would want someone to respect our ancestors, so we should respect the ancestors of others.

For Kevin, gene drives have imposed a godlike responsibility on us. 'As soon as you have the power, you have the responsibility, and you can't duck out of it.' But however good the intent, there is a danger that this path leads us to hubris. Kevin told me he wanted to 'create systems that are as beautiful' as evolution, but 'without so much suffering'. With gene drive, we could 'tame God's eight biblical plagues', alleviating famine by switching off the gene for swarming behaviour in desert locusts, or ease the suffering caused by a truly hellish parasite. In South America, screwworm larvae practise a truly diabolical life cycle: burrowing into the flesh of any animal with a wound as large as a tick bite, consuming living tissue as they go and emitting chemicals that summon adult female screwworms to lay their eggs in the wound – 'Until,' Kevin said, 'you have a macabre dance in which the screwworm is both conductor and performer, which very often ends up in the organism being devoured alive.'

Up to a billion South American animals each year suffer the excruciating attention of screwworms, he said. A gene drive could spread sterility throughout the screwworm population, wiping it from existence. 'How much suffering have we, as a species, caused in our time here? We could rebalance the ledger,' Kevin suggested, sparing some animals's suffering as 'a gesture of atonement for our species as a whole'; albeit, at the cost of the screwworm's extinction.

Rather than posing us as gods, a Māori-centred approach invites us to recognise that we, along with all living things, have inherited a thread of the divine. And this is the fundamental question presented by gene drive, and by gene-editing technologies in general: possessed of such tremendous power, what does it mean to be human? Driving our will through the genomes of other beings could set us further apart than we have ever been from the rest of life. But recognising the filament of divine life running through all things, as the Māori do, and making this the basis for the decisions we take, would draw tighter the threads that stitch us into the weave of the living world.

The Lion-Man shows us that the world can be reimagined, and that the basis of life is in relationship with others. These lessons can inform some of the most profound questions humanity has faced, in terms of our responsibility to the rest of creation. But, as Kevin's encounters with Māori make clear, the Lion-Man's last lesson is perhaps the most important: to listen.

With ears pricked, the Lion-Man is attentive to his surroundings.

His leap depends on what he hears. As we stand on the brink of a new chimerical age, we too ought to listen carefully, especially to those voices that have historically been ignored. For Māori, using synthetic biology responsibly means recognising that gene editing intervenes not just in a species' genome, but in its relationships with other living beings. And while we may not be able to ask the organisms whether they would choose to have their genomes engineered by us, there is still much we can learn by attending to the miraculous ways life engineers itself.

When I travelled to Boston to learn about bioplastics from Shannon Nangle and Nic Lee, I also paid a visit to Michael Levin's laboratory, across town at Tufts University. Michael was unable to see me, but I had arranged to meet his colleague, Doug Blackiston. Doug and Michael are part of a team of experts in developmental biology, robotics and AI whose research blurs the distinction between engineering and biology: they are building biological robots. Doug's role is to sculpt the robots, by hand, out of cell tissue harvested from *Xenopus laevis*, the African clawed frog.

The first xenobots, as the biological robots came to be called, were simple, comma-shaped machines designed to walk (in a pleasing coincidence, *xenopus* means 'strange foot'). The bodies were rough cubes of frog-skin cells, with two stubby legs made of heart muscle providing a crude motor. Following an AI-designed body plan, Doug handcrafted each robot in turn, each one less than a millimetre wide, coaxing the cells into shape with an ultra-fine stylus. The finished xenobots could indeed walk; but they could also do a number of other things the team hadn't anticipated. These organic automatons could work together to move particles around their environment, and unlike mechanical robots would self-heal when injured.

With the next generation, things got really weird. Xenobots

2.0 were formed from what is called an animal cap, stem cells extracted from frog embryos, and then allowed to develop independently, without relying on the algorithm. As is typical with an animal cap drawn from skin cells, the new xenobots grew hair-like cilia all over their bodies. But these hairs became more like limbs, flailing rapidly to allow the xenobot to swim through its environment. Rather than building a tadpole, the stem cells responded to the unique conditions of the laboratory environment to build what seemed to be an entirely new organism, a living tumbleweed – a feat no less strange than if a swarm of bees had emerged from a clump of frogspawn.

The second-generation xenobots seemed to behave with intent, navigating mazes or even signalling to each other by sending out pulses of calcium ions. Clusters of xenobots would apparently work together to 'tidy' particles and loose cells in their environment into neat piles. It only took a few days for the assembled cells to knit together and launch themselves into the world as new xenobots. The curious, house-proud behaviour was, Doug realised, in fact a form of crowd-sourced reproduction.

Xenobots 2.0 could typically make only one, at most two, generations of new xenobots before they began to decay. The team again consulted the evolutionary AI program, which predicted that a shape like a swollen 'C', or a pizza with one slice removed – essentially, Pac-Man – would be a more effective self-replicating machine. Hand-crafted again by Doug, Xenobots 3.0 had all the affordances of their forebears and could produce up to four generations of 'offspring'. Each generation has a limited lifespan. The bots are their own energy source; when the nutrients in their cells are exhausted, usually after a few weeks, they expire.

The handmade cells bear witness to life's incredible plasticity. Presented with a new environment, the *Xenopus laevis* stem cells rose to the challenge, unearthing new possibilities in their own body plans. Skin cells become swimming cells become self-replicating swarms of biological machines set on seeking out stray particles in their environment.

The Levin lab is located in suburban north Boston, in a modern, glass-fronted building huddled behind a two-storey car park. All around it are quiet tree-lined streets of clapboard houses, with yellow school buses passing by; to the north, the Mystic River winds its way towards Boston Harbor. At the entrance to the building, a three-metre-long polythene tube writhed blindly in the breeze, its tail wrapped around a bollard while its open mouth softly explored the kerb.

Doug Blackiston is tall, with an earnest, amenable manner. Once inside, he took me to the lab where he makes the xenobots: a small, plain room with a single microscope on a bench and a battered portable stereo. The first thing he showed me was the stylus he uses to hand-sculpt the animal caps. It looked impossibly fine, no larger than a scrap of electrical wire with two exposed filaments at the end. I couldn't imagine getting a proper grip on it, never mind using it to shape something so small.

You must be good at embroidery, I said. He agreed that it was a reassuringly old-fashioned process.

Doug picked up a circular culture dish with what looked like twenty or so tiny black seeds scattered in clear liquid. 'These are xenobots my student made, about a week ago,' he said, sliding the dish under the microscope.

I bent my eye towards the lens, and the black seeds erupted like popcorn into fluffy-looking orange spheres. Oh wow, I said. I couldn't help but laugh in delight. It was subtle, but they were definitely moving. Some drifted, others spun in a circle.

'This is like the clay that we use to build with,' Doug said.

THE LION-MAN'S LEAP

'I haven't done anything to modify them yet. Today's the very first day that the little hairs on their surface pop up. As the week progresses, they will begin to move faster.'

He produced another slide. 'So, these are the extreme senior citizens. They're near the end of their lifespan.' Where the first batch had been a deep orange, these were much paler, having lost much of their pigmentation as they burned through their energy stores. Despite their depleted state, they were actually moving more quickly. Most of their energy is in the form of lipids, Doug explained, which are heavy. The older xenobots' body mass was so low by now that it didn't take much to move them around.

Xenobots show that the Lion-Man's leap – from what is to what could be – is one that life makes constantly. In 2016, Japanese researchers discovered a microbe had evolved to use PET plastic as its main source of energy and carbon, in soil outside a plastic bottle recycling plant in Sakai. The natural function of *Ideonella sakaiensis* – a lozenge-shaped microbe with long, thread-like cilia – is to degrade cutin, the waxy coating that stops leaves from dehydrating. It attaches itself to the plastic and releases two enzymes, called PETases, which hydrolyse PET, converting it into the harmless monomers terephthalic acid and ethylene glycol. Presented with a new food source in the form of plastic, the microbe had found a way to exploit it in its ancient biochemistry.

One study has catalogued over 30,000 soil and ocean microbes with the potential to degrade ten types of common plastic. Extracting PETase from evolved microorganisms like *Ideonella sakaiensis* could help us clean up plastic waste, including the millions of tons fouling the world's oceans. Doug explained that the xenobots could be shaped to work with PETase and become filters that sense their environments. Xenobots that carry a particular protein glow green under certain wavelengths, but turn red when exposed to others, 'remembering' their exposure hours later. More sophisticated generations could be engineered to sense chemicals such as phthalates that are soaked into marine plastic and leach out into the surrounding water. The xenobots could be edited to emit a signal that shows humans where to deploy enzymes like PETase. They could even become a PETase delivery system, engineered to seek out the most contaminated particles and round them up, then discharge the enzyme themselves. Once the food stored in their cells was exhausted, the xenobots would simply perish in a bath of harmless chemical monomers.

Xenobots could also be used to deliver bioremediating agents to clean up other forms of pollution in other kinds of environments, harnessing the evolutionary potential of microbes to clean up some of our worst messes. Lanmodulin, a protein produced by a common soil microbe, binds tightly to americium and curium, two of the most toxic metals produced in nuclear fission. Much of the 3 million barrels of oil spilled during the Deepwater Horizon disaster was bioremediated by applying microbes that have known how to break down hydrocarbons for millennia, and a PCB-degrading bacterium called *Dehalococcoides* has been used to reduce the toxicity of sediments in the Hudson River.

We sat in silence for a moment as I watched the xenobots drift and flow in their meandering, uniquely xenobot way. Individual kernels spun in tight circles. Pairs slow-danced a pas de deux while others collided and rebounded, leaping together and apart.

It is so peaceful, I said. 'It's very Zen, yeah,' Doug agreed.

It may be inevitable that we have to edit or engineer nature to cope with the world we have made. The Lion-Man's leap could be the best hope for our harrowed, chimerical world. But the Lion-Man doesn't only leap; he listens. When we listen to what other species know, whole worlds of possibility crack open. The microbial world is a riot of love and theft, living communally, freely sharing their DNA, and forming, breaking and repairing alliances with a breathtaking level of realpolitik. Xenobots show that life is not a plan, more a kind of embodied possibility. In these instances, what a microbe or a xenobot knows is the most essential truth: that what we are does not limit what we could be.

One evening in December 2005, twin sisters Margaret and Christine Wertheim met at Christine's house in Queensland to spend the evening crocheting, as they had done together since childhood. On this occasion, however, they began to experiment. It is a quirk of science that following a conventional crochet formula of 'stitch n, increase 1' produces mathematically perfect hyperbolic geometry – a geometry of negative curvature, in which parallel lines can converge and diverge. Hyperbolic shapes are fiendishly difficult to visualise, but they are everywhere in nature. The frills on lettuce leaves are hyperbolic, as are the roiling, curling forms of coral (hyperbolic shapes maximise surface area, which means the polyps and their symbionts have the greatest possible opportunity to draw down nutrients). Even the humble sea slug knows geometry: the flanges on nudibranchs have negative curvature.

By varying the regular rate of increase – instead of crocheting three stitches then increasing one stitch, they would crochet five then increase one, or seven then one – Margaret and Christine discovered that their work looked a lot less like perfect mathematical models, and more like living things; specifically, they looked a lot like coral reefs, which scientists had just begun to warn were at risk from warming temperatures. That first small woollen reef on the Wertheims' table has since grown into what may be the world's largest collaborative art project, incorporating fifty-two 'satellite' reefs across the world, and involving around 25,000 people, almost all of them women, and between 400,000 and 500,000 hours of work by hand.

The process is disarmingly simple – take the formula and vary it – but the results are astonishing. Every type of coral is featured in the hyperbolic Crochet Coral Reef, from branching species like staghorn and elkhorn to fingery sun coral and feathery carnation coral. Mushroom-domed bubble coral and the mazy globes of brain coral sit alongside an array of tubes, plates and rosettes. The scale is vast, with some reefs taking up hundreds of square feet of gallery space; others would sit in the palm of your hand. The sheer abundance of shade and shape on display, and the wired, paint-bomb colour schemes, is electrifying; all the more so when contrasted with the bleached and ashen reality. Some contributors have conjured entirely new coral species by experimenting with the infinite fund of forms available in the simple hyperbolic code. Others bring in different materials, swapping wool for cotton or silk – even plastic shopping bags and video tape. Each variation has synthesised a new kind of coral organism. 'We began to realise, this is like evolution,' Margaret told me. 'From this very simple formula you can then complexify and diversify, just like DNA complexifies and diversifies, to an endless, never-exhausted taxonomy of crochet coral beings.'

It barely needs saying that the work of women's hands has been belittled or neglected for much of history. By casting coral in wool, the crochet reef asks questions about our similarly cavalier attitude to the work of nature. Of the fifty-two satellite reefs assembled and exhibited around the world so far, only three still exist, Margaret told me ruefully. Museums and galleries didn't want the trouble of finding somewhere to store the work; in each case, following an exhibition almost the entire handmade reef had been thrown away.

The handcrafted nature of these synthetic reefs also recalls the very hands-on approach to assisting in coral's evolution: manually transplanting corals from one place to another, or priming coral spawning with buckets and pipettes. Crocheting coral entails quite literally developing a feel for what coral 'knows', and every stitch deepens an appreciation for their geometric genius. Like the makers of the Lion-Man, patiently carving their chimera from a piece of mammoth ivory, this slow investment of time and touch is also a honing of the imagination. If we are to lend a hand in the evolution of other species, we need to gain a feel for the infinite potential for change in living things; for making and for repair, one stitch at a time. The Crochet Coral Reef takes us back to that dark cave, 40,000 years ago, where hands and minds were just beginning to imagine what kinds of transformation might be possible.

KAFKA'S LEOPARDS

The old ways are broken and new ones have yet to take shape. But a better way of life for all is possible, if we pay attention to what the natural world can teach us. One evening in November 1917, Franz Kafka retired to bed and added the briefest of parables to his notebook:

> Leopards invade the Temple and drain the sacred vessels; this happens again and again; in time, it becomes predictable, and a part of the ritual.

Kafka doesn't say what drives the leopards into the temple. It may be drought, or a new road, or farmers burning the forest for pasture. But what comes next changes everything: born of the need to live together, a new way of being in which all life can thrive becomes possible; and the animals show the way.

Nature's genius is blessed by an irrepressible creativity; even under the most stringent pressures, evolution's abundant inspiration is finding a way; trammelled and travestied it may be, but life does not stop searching for the new shape to fit the new conditions. In doing so, it shows how we too can change; that, in fact, we must. Evolution's ingenuity alone won't be enough. Experimenting under pressure will lead life down many false avenues, sometimes fatally. Engineering nature ourselves may offer a vital helping hand, but the only real hope for many species lies in our capacity and willingness to transform our whole way of life. Fortunately, that capacity has always been in us. We harbour great wells of plasticity: by drawing from them, we could remake our cities and reimagine everyday life, from the materials we use to the food we eat and the whole basis of our economies; we could synchronise with nature's rhythms and find kinship across the unpassable gulf of animal Umwelts. All this vitality is in us, too, rooted deep in the essence of what it means to be human.

In Kafka's parable the animals take the first step, but the leopards survive because the priests learn to keep replenishing the pitchers. We have always been more than we think, that much is certain; the question is, are we willing to embrace it?

If we did, what else might we become?

LIST OF CITED SOURCES

Introduction

Brondizio, E. S., et al. *Global Assessment Report on Biodiversity and Ecosystem Services of the Intergovernmental Science-Policy Platform on Biodiversity and Ecosystem Services*. IPBES, 2019.

Brown, Charles, and Mary Bomberger Brown. 'Where Has All the Road Kill Gone?' *Current Biology* 23/6 (2013).

Ghadouani, Anas, and Bernadette Pinel-Alloul. 'Phenotypic Plasticity in Daphnia pulicaria'. *Journal of Plankton Research* 24/10 (2002).

Harvard Medical School. 'The Evolution of Bacteria on a "Mega-Plate" Petri Dish (Kishony Lab), at: <https://youtu.be/plVk4NVIUh8?si=lE-f4JPVvm7F-ZQN>.

Keller, Evelyn Fox. *The Century of the Gene*. Harvard University Press, 2000.

Mukherjee, Siddhartha. *The Gene: An Intimate History*. Vintage, 2017.

Oostra, Vicencio, et al. 'Strong Phenotypic Plasticity Limits Potential for Evolutionary Responses to Climate Change'. *Nature Communications* 9/1005 (2018).

Otto, Sarah. 'Adaptation, Speciation, and Extinction in the Anthropocene'. *Proceedings of the Royal Society B* 285 (2018).

Palovacs, Eric, et al. 'Fates Beyond Traits: Ecological Consequences of Human-Induced Trait Change'. *Evolutionary Applications* 5 (2012).

Palumbi, Stephen. 'Humans as the World's Greatest Evolutionary Force'. *Science* 293 (2001).

Rodewald, Amanda, et al. 'Dynamic Selective Environments and Evolutionary Traps in Human-Dominated Landscapes'. *Ecology* 92/9 (2011).

Ruthsatz, Katharina, et al. 'Microplastics Ingestion Induces Plasticity in Digestive Morphology in Larvae of *Xenopus laevis*'. *Comparative Biochemistry and Physiology A* 269 (2022).

Scheffers, Brett R., et al. 'The Broad Footprint of Climate Change from Genes to Biology to People'. *Science* 354 (2016).

Standen, Emily, et al. 'Developmental Plasticity and the Origin of Tetrapods'. *Nature* 513 (2014).

Traill, Lochran, et al. 'Demography, Not Inheritance, Drives Phenotypic Change in Hunted Bighorn Sheep'. *PNAS* 111/36 (2014).

Walsh, D. M. *Organism, Agency and Evolution*. Cambridge University Press, 2015.

1: *Optimum Dog*

Agnvall, Beatrix, et al. 'Is Domestication Driven by Reduced Fear of Humans?' *Biology Letters* 11 (2015).

Bennett, Cary, et al. 'The Broiler Chicken as a Signal of Human Reconfigured Biosphere'. *Royal Society Open Science* 5 (2015).

Browne, Janet. *Charles Darwin*. Pimlico, 2002.

Capildeo, Vahni, *Venus as a Bear*. Carcanet, 2018.

Chirchir, Habiba, et al. 'Recent Origin of Low Trabecular Bone Density in Modern Humans'. *PNAS* 112/2 (2014).

LIST OF CITED SOURCES

Curry, Andrew. 'Archaeology: The Milk Revolution'. *Nature* 500 (2013).

Darwin, Charles. *The Variation of Animals and Plants under Domestication.* John Murray, 1868.

Dugatkin, Lee Alan, and Lyudmila Trut. *How to Tame a Fox.* University of Chicago Press, 2017.

Fish, Kenneth. *Living Factories: Biotechnology and the Unique Nature of Capitalism.* McGill-Queens University Press, 2013.

Gale, Edwin. *The Species That Changed Itself.* Penguin, 2020.

Guthman, Julie. 'The CAFO in the Bioreactor'. *Environmental Humanities* 14/1 (2022).

Husain, Amber. *Meat Love: An Ideology of the Flesh.* Mack, 2023.

Irving-Pease, Evan, et al. 'Paleogenomics of Animal Domestication'. In: Charlotte Lindqvist and Om Rajora (eds), *Paleogenomics.* Springer, 2018.

Kennis & Kennis, at: <https://www.kenniskennis.com>.

Lispector, Clarice. *Near to the Wild Heart.* Penguin, 2014.

MacHugh, Daniel, et al. 'Taming the Past: Ancient DNA and the Study of Animal Domestication'. *Annual Review of Animal Biosciences* 5 (2017).

Metcalf, Jacob. 'Meet Shmeat: Food System Ethics, Biotechnology and Re-Worlding Technoscience'. *Parallax* 19/1 (2013).

Monbiot, George. *Regenesis: Feeding the World Without Devouring the Planet.* Allen Lane, 2022.

Pendleton, Amanda, et al. 'Comparison of Village Dog and Wolf Genomes Highlights the Role of the Neural Crest in Dog Domestication'. *BMC Biology* 16/64 (2018).

Price, Michael. 'Early Humans Domesticated Themselves'. *Science* (4 December 2019).

Protsiv, Myroslava, et al. 'Decreasing Human Body Temperature in the US Since the Industrial Revolution'. *eLife* 9 (2020).

Ryan, Timothy, and Colin Shaw. 'Gracility of the Modern Homo sapiens Skeleton is the Result of Decreased Biochemical Loading'. *PNAS* 112/2 (2014).

Seba, Tony, and Catherine Tubb. *Rethinking Food and Agriculture 2020–2030*. ReThinkX, 2020.

Tixier-Boichard, Michèle, et al. 'Chicken Domestication: From Archaeology to Genomics'. *C. R. Biologies* 334 (2011).

Wrangham, Richard. *The Goodness Paradox*. Pantheon Books, 2019.

2: The Living City

Alberti, Marina, et al. 'Urban Driven Phenotypic Changes'. *Philosophical Transactions of the Royal Society B* 372 (2017).

Altermatt, Florian, and Dieter Ebert. 'Reduced Flight-to-Light Behaviour of Moth Populations Exposed to Long-Term Urban Light Pollution'. *Biology Letters* 12 (2016).

Armstrong, Rachel. *Soft Living Architecture*. Bloomsbury, 2018.

Badyaev, Alexander. 'Evolution on a Local Scale: Developmental, Functional, and Genetic Bases of Divergence in Bill Form and Associated Changes in Song Structure Between Adjacent Habitats'. *Evolution* 62/8 (2008).

Barua, Maan. *Living Cities: Reconfiguring Urban Ecology*. University of Minnesota Press, 2023.

Benyus, Janine. 'The Generous City'. *Architectural Design* 85/4 (2015).

Benyus, Janine, et al. 'Ecological Performance Standards for Regenerative Urban Design'. *Sustainability Science* 17/6 (2022).

Biomimicry 3.8, at: <https://www.biomimicry.net>.

Certeau, Michel de. *The Practice of Everyday Life*. University of California Press, 1984.

Cheptou, Pierre-Olivier, et al. 'Adaptation to Fragmentation'. *Philosophical Transactions of the Royal Society B* 372 (2017).

Crissman, J. R., et al. 'Population Genetic Structure of the German Cockroach (*Blattodea: Blattellidae*) in Apartment Buildings'. *Journal of Medical Entomology* 47/4 (2010).

Dahirel, Maxime, et al. 'Urbanization-Driven Changes in Web Building and Body Size in an Orb Web Spider'. *Journal of Animal Ecology* 88/1 (2019).

Datta, Ayona. 'India's Ecocity? Environment, Urbanisation and Mobility in the Making of Lavasa'. *Environment and Planning C: Politics and Space* 30 (2012).

Davis, Anthony, and Thomas Glick. 'Urban Ecosystems and Island Biogeography'. *Environmental Conservation* 5/4 (1978).

DeLillo, Don, *Underworld*. Scribner, 1997.

Dunn, Robert. 'A Theory of City Biogeography and the Origin of Urban Species'. *Frontiers in Conservation Science* 3 (2022).

Ferracina, Simone. *Ecologies of Inception*. Routledge, 2021.

Fusco, N. A., et al. 'Urbanization Reduces Gene Flow but not Genetic Diversity of Stream Salamander Populations in the New York City Metropolitan Area'. *Evolutionary Applications* 14/1 (2020).

Gomes, D. G. E. 'Orb-Weaving Spiders are Fewer but Larger and Catch More Prey in Lit Bridge Panels from a Natural Artificial Light Experiment'. *PeerJ* 17/8 (2020).

Gould, Stephen Jay, and Elizabeth Vrba. 'Exaptation – a Missing Term in the Science of Form'. *Paleobiology* 8.1 (1982).

Hiemstra, Auke-Florian, et al. 'Bird Nests Made from Anti-bird Spikes'. *Deinsea* 21(2023).

Ichioka, Sarah, and Michael Pawlyn. *Flourish*. Triarchy Press, 2022.

Jemisin, N. K. *The City We Became*. Little, Brown, 2020.

Jencks, Charles, and Nathan Silver. *Adhocism*. MIT Press, 2013.

Kern, E. M. A., and R. B. Langerhans. 'Urbanization Alters Swimming Performance of a Stream Fish'. *Frontiers in Ecology and Evolution* 6 (2019).

Kerstes, N. A. G., et al. 'Snail Shell Colour Evolution in Urban Heat Islands Detected via Citizen Science'. *Communications Biology* 2/264 (2019).

Lescak, Emily, et al. 'Evolution of Stickleback in 50 years on Earthquake-uplifted Islands'. *PNAS* 112/52 (2015).

Liker, András. 'Adaptive Changes in Urban Populations'. *Biologia Futura* 71 (2020).

Littleford-Colquhoun, Bethan, et al. 'Archipelagos of the Anthropocene'. *Molecular Ecology* 26 (2017).

Melis, Alessandro, and Telmo Pievani. 'Exaptation as a Design Strategy for Resilient Communities'. In N. Rezaei (ed.), *Transdisciplinarity*. Springer, 2022.

Munshi-South, J., and K. Kharchenko. 'Rapid, Pervasive Genetic Differentiation of Urban White-Footed Mouse (*Peromyscus leucopus*) Populations in New York City'. *Molecular Ecology* 1 (2010).

Potvin, Dominique, and Kirsten Parris. 'Song Convergence in Multiple Urban Populations of Silvereyes (*Zosterops lateralis*)'. *Ecology and Evolution* 2/8 (2012).

Putman, B. J., et al. 'Downsizing for Downtown: Limb Lengths, Toe Lengths, and Scale Counts Decrease with Urbanization in Western Fence Lizards (*Sceloporus occidentalis*)'. *Urban Ecosystems* 22 (2019).

Roberts, Michael Symmons, and Paul Farley. *Edgelands*. Random House, 2011.

Schilthuizen, Menno. *Darwin Comes to Town*. Quercus, 2018.

Schmidt, Chloé, and Colin Garroway. 'Systemic Racism Alters Wildlife Genetic Diversity'. *PNAS* 119/43 (2022).

Sierro, J., et al. 'European Blackbirds Exposed to Aircraft Noise Advance Their Chorus, Modify Their Song and Spend More Time Singing'. *Frontiers in Ecology and Evolution* 5/68 (2017).

Solnit, Rebecca. *Whose Story Is This?* Granta, 2019.

Van Dongen, W. F. D., et al. 'Variation at the DRD4 Locus is Associated with Wariness and Local Site Selection in Urban Black Swans'. *BMC Evolutionary Biology* 15 (2015).

Vince, Gaia. *Nomad Century*. Penguin, 2023.

Watnick, Paula, and Roberto Kolter. 'Biofilm, City of Microbes'. *Journal of Bacteriology* 182/10 (2000).

Wilson, Edward O. *The Diversity of Life*. Penguin, 1992.

3: One Touch Makes the Whole World Kin

Ah-King, Malin, and Eva Hayward. 'Toxic Sexes: Perverting Pollution and Queering Hormone Disruption'. *Technosphere Magazine* (March 2019).

Benjamin, Walter. *One-Way Street*. Penguin, 2009.

Chen, Mel. 'Toxic Animacies, Inanimate Affections'. *GLQ* 17/2–3 (2011).

Crews, David, and John MacLachlan. 'Epigenetics, Evolution, Endocrine Disruptors, Health and Disease'. *Endocrinology* 147/6 (2006).

Daley, Jason. 'Science Is Falling Woefully Behind in Testing New Chemicals'. *Smithsonian Magazine* (3 February 2017).

Di Chiro, Giovanna. 'Polluted Politics?' In: Catriona Mortimer-Sandilands and Bruce Erickson (eds), *Queer Ecologies*. Indiana University Press, 2010.

Dickinson, Adam. *Anatomic*. Coach House Books, 2018.

Dickinson, Adam. 'Metabolic Pathways'. *Jacket2* (21 June 2019). At: <https://jacket2.org/commentary/metabolic-pathways>.

Goiran, Claire, et al. 'Industrial Melanism in the Seasnake Emydocephalus annulatus'. *Current Biology* 27/16 (2017).

Hairston, N., et al. 'Rapid Evolution Revealed by Dormant Eggs'. *Nature* 401 (1999).

Hird, Myra. 'Animal Transex'. *Australian Feminist Studies* 21/49 (2006).

Homer. *The Odyssey*. Translated by Emily Wilson. W. W. Norton & Co., 2017.

Kirksey, Eben. 'Chemosociality in Multispecies Worlds'. *Environmental Humanities* 12/1 (2020).

Langston, Nancy. 'Endocrine Disruptors in the Environment'. In: Daniel Lee Kleinmann et al. (eds), *Controversies in Science and Technology*. Oxford University Press, 2010.

Meikle, Jeffrey. *American Plastic*. Rutgers University Press, 1997.

Murphy, Michelle. 'Chemical Infrastructure of the St. Clair River'. In: Soraya Boudia and Nathalie Jas (eds), *Toxicants, Health and Regulation Since 1945*. Pickering and Chatto, 2013.

Murphy, Michelle. 'Alterlife and Decolonial Chemical Relations'. *Cultural Anthropology* 32/4 (2017).

Ogasa, Nikk. 'Soil Microbe Could Clean Up Nuclear Waste'. *Scientific American* 325 (2021).

Oxman. At: <https://oxman.com>.

Pessoa, Fernando. *The Book of Disquiet*. Profile, 2017.

Pradham, S., et al. 'Nature-Derived Materials for the Fabrication of Functional Biodevices'. *Materials Today Bio* 7 (2020).

Preciado, Paul. *Can the Monster Speak? Report to an Academy of Psychoanalysis*. Fitzcarraldo, 2021.

Reid, Noah. 'The Genomic Landscape of Rapid Repeated Evolutionary Adaptation to Toxic Pollution in Wild Fish'. *Science* 354 (2016).

Rosenboom, Jan-Georg, et al. 'Bioplastics for a Circular Economy'. *Nature Reviews: Materials* 7 (2022).

Shadaan, Reena, et al. 'Endocrine-Disrupting Chemicals (EDCs) as Industrial and Settler Colonial Structures'. *Catalyst* 6/1 (2020).

Tokarczuk, Olga. *Flights*. Fitzcarraldo, 2007.
Toro-Valdivieso, Constanza, et al. 'Heavy Metal Contamination in Pristine Environments: Lessons from the Juan Fernandez Fur Seal (*Arctocephalus philippii philippii*)'. *Royal Society Open Science* 10/3 (2023).
Vandenberg, Laura, et al. 'Plastic Bodies in a Plastic World'. *Journal of Cleaner Production* 140 (2017).
Van't Hof, A. E., et al. 'The Industrial Melanism Mutation in British Peppered Moths is a Transposable Element'. *Nature* 534 (2016).
Wei Ren, et al. 'Possibilities and Limitations of Biotechnical Plastic Degradation and Recycling'. *Nature Catalysis* 3 (2020).
Wirgin, Isaac. 'Mechanistic Basis of Resistance to PCBs in Atlantic Tomcod from the Hudson River'. *Science* 331 (2011).

4: The Kinship of Languages

Allen, Jenny, et al. 'Cultural Revolutions Reduce Complexity in the Songs of Humpback Whales'. *Proceedings of the Royal Society B* 285 (2018).
Andreas, Jacob, et al. 'Toward Understanding the Communication of Sperm Whales'. *iScience* 25 (2022).
Atwood, Margaret. *Eating Fire: Selected Poetry 1965–1995*. Virago, 1998.
Ball, Philip. 'The Challenges of Animal Translation'. *New Yorker* (27 April 2021).
Benjamin, Walter. *Illuminations*. Harcourt Brace Jovanovich, 1968.
Blair, Hannah, et al. 'Evidence for Shipping Noise Impacts on Humpback Whale Foraging Behaviour'. *Biology Letters* 12 (2016).

Boroditsky, Lera, and Alice Gaby. 'Remembrances of Times East: Absolute Spatial Representations of Time in an Australian Aboriginal Community'. *Psychological Science* 21/11 (2010).

Budiansky, Stephen. *If a Lion Could Talk*. Weidenfeld and Nicholson, 1998.

Carr, Holly Corfield. *Subsongs*. National Trust, 2018.

Chase, Ava. 'Music Discriminations by Carp (*Cyprinus carpio*)'. *Animal Learning and Behaviour* 29/4 (2001).

Chiang, Ted. *Stories of Your Life and Others*. Picador, 2015.

Crates, Ross, et al. 'Loss of Vocal Culture and Fitness Cost in a Critically Endangered Songbird'. *Proceedings of the Royal Society B* 288 (2021).

Daria, Charlie, et al. 'Effects of Anthropogenic Noise on Cognition, Bill Colour, and Growth in the Zebra Finch (*Taeniopygia guttata*)'. *Acta Ecologica* 26.3 (2023).

Dieguez, Sebastian, and Julien Bogousslavsky. 'Baudelaire's Aphasia: From Poetry to Cursing'. *Frontiers of Neurological Neuroscience* 22 (2007).

Dunlop, Rebecca. 'The Effects of Vessel Noise on the Communication Network of Humpback Whales'. *Proceedings of the Royal Society B* 6 (2019).

Erbe, Christine, et al. 'The Effects of Ship Noise on Marine Mammals'. *Frontiers in Marine Science* 6 (2019).

Fitch, W. Tecumseh. 'The Biology and Evolution of Music'. *Cognition* 100 (2006).

Fitch, W. Tecumseh. 'Musical Protolanguage'. In Johan Bolhuis and Martin Everaert (eds), *Birdsong, Speech and Language*. MIT Press, 2016.

Fournet, Michelle. 'Humpback Whales Alter Calling Behaviour in Response to Natural Sounds and Vessel Noise'. *MEPS* 607 (2018).

LIST OF CITED SOURCES

Frost, Robert. *The Collected Poems*. Penguin, 2013.

Garland, Ellen, et al. 'When Does Cultural Evolution Become Cumulative Culture?' *Philosophical Transactions of the Royal Society B* 377 (2021).

Garland, Ellen, and Peter MacGregor. 'Cultural Transmission, Evolution and Revolution in Vocal Displays'. *Frontiers in Psychology* 11 (2020).

Gordon, Timothy, et al. 'Habitat Degradation Negatively Effects Auditory Settlement Behaviour in Coral Reef Fishes'. *PNAS* 115/20 (2018).

Gordon, Timothy, et al. 'Acoustic Enrichment Can Enhance Fish Community Development on Degraded Coral Reef Habitat'. *Nature Communications* 10 (2019).

Heller-Roazen, Daniel. *Echolalias*. Zone Books, 2005.

Hoeschele, Marisa, et al. 'Searching for Origins of Musicality Across Species'. *Philosophical Transactions of the Royal Society B* 370 (2015).

Honing, Henkjan. *The Origins of Musicality*. MIT Press, 2019.

Honing, Henkjan. *The Evolving Animal Orchestra*. MIT Press, 2019.

Iniesta, Iván. 'Tomas Tranströmer's Stroke of Genius'. *Progress in Brain Research* 206 (2013).

Juárez, Roselvy, et al. 'House Wrens *Troglodytes aedon* Reduce Repertoire Size and Change Song Element Frequencies in Response to Anthropogenic Noise'. *IBIS* 163 (2021).

'Koko the Gorilla – Message for Humans'. At: <https://youtu.be/cfj109kYgzw?si=7PkJ4ExXU6B3P-xI>.

Krause, Bernie, and Almo Farina. 'Using Ecoacoustic Methods to Survey the Impacts of Climate Change on Biodiversity'. *Biological Conservation* 195 (2018).

Laxness, Halldór, *Fish Can Sing*. Penguin, 2022.

Loud Numbers. At: <https://www.loudnumbers.net>.

McKay, Laura Jean. *The Animals in That Country*. Scribe, 2021.
Melville, Herman. *Moby-Dick; or, The Whale*. Penguin, 2013.
Meyer, Julien. 'Environmental Linguistic Typology of Whistled Languages'. *Annual Review of Linguistics* 7 (2021).
Morrison, C. A., et al. 'Bird Population Declines and Species Turnover are Changing the Acoustic Properties of Spring Soundscapes'. *Nature Communications* 12 (2021).
Mullender, Rob. 'Divine Agency: Bringing to Light the Voice Figures of Margaret Watts-Hughes'. *Sound Effects* 8/1 (2019).
Murray, Les. *New Collected Poems*. Carcanet, 2003.
Nabokov, Vladimir. *Speak, Memory*. Penguin, 2016.
Nordlinger, Rachel, et al. 'Sentence Planning and Production in Murrinhpatha, an Australian "Free Word Order" Language'. *Linguistic Society of America* 98/2 (2022).
Osbrink, Alison, et al. 'Traffic Noise Inhibits Cognition Performance in a Songbird'. *Proceedings of the Royal Society B* 288 (2021).
Otter, Ken. 'Continent-Wide Shifts in Song Dialects of White-throated Sparrows'. *Current Biology* 30 (2020).
Owen, Clare, et al. 'Migratory Convergence Facilitates Cultural Transmission of Humpback Whale Song'. *Royal Society Open Science* 6 (2019).
Porter, Debra, and Allen Neuringer. 'Music Discrimination by Pigeons'. *Journal of Experimental Psychology* 10/2 (1984).
Potvin, Dominique, and Scott MacDougal-Shackleton. 'Traffic Noise Affects Embryo Mortality and Nesting Growth Rates in Captive Zebra Finches'. *Journal of Experimental Zoology* 323 (2015).
Prather, Jonathan, et al. 'Brains for Birds and Babies'. *Neuroscience and Biobehavioral Reviews* 81 (2017).
Project CETI. At: <https://www.projectceti.org/>.
Rekdahl, Melinda, et al. 'Culturally Transmitted Song Exchange

Between Humpback Whales in Southeast Atlantic and Southeast Indian Ocean Basin'. *Royal Society Open Science* 5 (2018).

Risch, Denise, et al. 'Changes in Humpback Whale Song Occurrence in Response to an Acoustic Source 200 Kilometres Away'. *PLOS One* 7.1 (2012).

Rohrmeier, Martin, et al. 'Principles of Structure Building in Music, Language and Animal Song'. *Philosophical Transactions of the Royal Society B* 370 (2015).

Rutz, Christian, et al. 'Using Machine Learning to Decode Animal Communication'. *Science* 381 (2023).

Sacks, Oliver, *Musicophilia*. Picador, 2011.

Schulze, Josephine, et al. 'Humpback Whale Song Revolutions Continue to Spread from the Central to the Eastern South Pacific'. *Royal Society Open Science* 9 (2022).

Simonis, Anne, et al. 'Co-occurrence of Beaked Whale Strandings and Naval Sonar in the Mariana Islands, Western Pacific'. *Proceedings of the Royal Society B* 287 (2020).

Stewart, Susan. 'Rhyme and Freedom'. In: Marjorie Perloff and Craig Dworkin (eds), *The Sound of Poetry/The Poetry of Sound*. Chicago University Press, 2009.

Tawada, Yoko. 'The Art of Being Nonsynchronous'. In: Marjorie Perloff and Craig Dworkin (eds), *The Sound of Poetry/The Poetry of Sound*. Chicago University Press, 2009.

Thomas, James, and Simon Kirkby. 'Self-Domestication and the Evolution of Language'. *Biological Philosophy* 33/9 (2018).

Thompson, Kirsten, et al. 'Urgent Assessment Needed to Evaluate Potential Impacts on Cetaceans from Deep Seabed Mining'. *Frontiers in Marine Science* 10 (2023).

Tranströmer, Tomas. *New Selected Poems*. Bloodaxe, 2011.

Trumble, Stephen, et al. 'Baleen Whale Cortisol Levels Reveal a Physiological Response to Twentieth Century Whaling'. *Nature Communications* 9 (2018).

Tsujii, Koki, et al. 'Change in Singing Behaviour of Humpback Whales Caused by Shipping Noise'. *PLOS One* 13/10 (2018).

Uexküll, Jakob von. *A Foray into the Worlds of Animals and Men.* Minnesota University Press, 2010.

Warren, Victoria, et al. 'Migratory Insights from Singing Humpback Whales Recorded Around Central New Zealand'. *Royal Society Open Science* 7 (2020).

Wiggins, Gerraint, et al. 'The Evolutionary Role of Creativity'. *Philosophical Transactions of the Royal Society B* 370 (2015).

Williams, Heather. 'Cumulative Cultural Evolution and Mechanisms for Cultural Selection in Wild Bird Songs'. *Nature Communications* 13 (2022).

Wolfenden, Andrew, et al. 'Aircraft Sound Exposure Leads to Song Frequency Decline and Elevated Aggression in Wild Chiffchaffs'. *Journal of Animal Ecology* 88 (2019).

Yong, Ed. *An Immense World.* The Bodley Head, 2022.

Zandberg, Lies, et al. 'Global Cultural Evolution Model of Humpback Whale Song'. *Philosophical Transactions of the Royal Society B* 376 (2021).

5: Strange Minds

Ananthaswamy, Anil. 'Heavy Metal Poisoning may be Changing Birds' Personalities'. *New Scientist* (22 March 2018).

Ashur, Molly, et al. 'Impacts of Ocean Acidification on Sensory Function in Marine Organisms'. *Integrative and Comparative Biology* 57/1 (2017).

Baraniuk, Chris. 'Birds "Dream Sing" by Moving their Vocal Muscles in their Sleep'. *New Scientist* (9 February 2018).

Blamires, S. J., and W. I. Sellers. 'Modelling Temperature and

LIST OF CITED SOURCES

Humidity Effects on Web Performance'. *Conservation Physiology* 7 (2019).

Bollier, David, and Silke Helfrich. *Free, Fair and Alive*. New Society Publishers, 2019.

Borges, Jorge Luis. *The Book of Imaginary Beings*. Penguin, 1974.

Brumm, Adam, et al. 'Oldest Cave Art Found in Sulawesi'. *Science Advances* 7/3 (2021).

Buolamwini, Joy. 'When the Robot Doesn't See Dark Skin'. *New York Times* (21 June 2018).

Buolamwini, Joy. 'Artificial Intelligence Has a Problem with Gender and Racial Bias'. *Time* (7 February 2019).

Calvo, Paco, et al. 'Plants are Intelligent, Here's How'. *Annals of Botany* 125 (2020).

Clark, Andy, and David Chalmers. 'The Extended Mind'. *Analysis* 58 (1998).

Crawford, Kate, and Trevor Paglen. 'Excavating AI'. At: <https://excavating.ai>.

Croney, Candace, and Sarah Boysen. 'Acquisition of a Joystick-Operated Video Task by Pigs (*Sus scrofa*)'. *Frontiers in Psychology* 12 (2021).

DeVore, Janya, et al. 'The Evolution of Targeted Cannibalism and Cannibal-induced Defences in Invasive Populations of Cane Toads'. *PNAS* 118/35 (2021).

Dillard, Annie. *Teaching a Stone to Talk*. Canongate, 2017.

Epstein, Ziv, et al. 'Interpolating GANs to Scaffold Autotelic Creativity'. *Joint Proceedings of the ICCC 2020 Workshops* (September 7–11 2020).

Facchin, Marco, and Giulia Leonetta. 'Extended Animal Cognition'. *Synthese* 203/5 (2024).

Fraser, Nancy. *Cannibal Capitalism*. Verso, 2022.

Godfrey-Smith, Peter. *Other Minds: The Octopus, the Sea, and the Deep Origins of Consciousness*. William Collins, 2017.

Godfrey-Smith, Peter. *Metazoa: Animal Minds and the Birth of Consciousness*. William Collins, 2020.

Grunst, Andrea, et al. 'Variation in Personality Traits Across a Metal Pollution Gradient in a Free-Living Songbird'. *Science of the Total Environment* 630 (2018).

Hickel, Jason. 'What Does DeGrowth Mean? A Few Points of Clarification'. *Globalisation* 18/7 (2015).

Hickel, Jason, et al. 'Degrowth Can Work – Here's How Science Can Help'. *Nature* 615 (2022).

Hildyard, Daisy. *The Second Body*. Fitzcarraldo, 2017.

Hoel, Erik. 'The Overfitted Brain: Dreams Evolved to Assist Generalization'. *Patterns* 2 (2021).

Japyassú, Hilton, and Kevin Laland. 'Extended Spider Cognition'. *Animal Cognition* 20 (2017).

Jelbert, Sarah, et al. 'New Caledonian Crows Infer the Weight of Objects from Observing their Movements in a Breeze'. *Proceedings of the Royal Society B* 286 (2019).

Johannesson, Jessica Gaitán. *The Nerves and Their Endings*. Scribe, 2021.

Kohn, Eduardo. *How Forests Think*. University of California Press, 2013.

Kohn, Eduardo. 'Forest Forms and Ethical Life'. *Environmental Humanities* 14/2 (2022).

Krause, Mark. 'Is a Murmuration of Birds a Conscious Thing?' *Medium* (25 September 2020).

Leca, Jean-Baptiste, et al. 'Acquisition of Object-robbing and Object/Food-Bartering Behaviours'. *Philosophical Transactions of the Royal Society B* 376 (2020).

Levin, Michael. 'The Computational Boundaries of a "Self"'. *Frontiers in Psychology* 10 (2019).

Levin, Michael. 'Life, Death and the Self'. *Biochemical and Biophysical Research Communications* 564 (2021).

LIST OF CITED SOURCES

Little, Alexander, et al. 'Population Differences in Aggression are Shaped by Tropical Cyclone-Induced Selection'. *Nature Ecology and Evolution* 3 (2019).

Livingston, Julie. *Self-Devouring Growth.* Duke University Press, 2019.

Malinowski, J. E., et al. 'Do Animals Dream?' *Consciousness and Cognition* 95 (2021).

Milward-Hopkins, Joel, et al. 'Providing Decent Living with Minimum Energy: A Global Scenario'. *Global Environmental Change* 65 (2020).

Nagelkerken, Ivan, and Philip Mundy. 'Animal Behaviour Shapes the Ecological Effects of Ocean Acidification and Warming'. *Global Change Biology* 22 (2016).

Nixon, Rob. 'The Less Selfish Gene'. *Environmental Humanities* 13/2 (2021).

Osbrink, Alison, et al. 'Traffic Noise Inhibits Cognition Performance in a Songbird'. *Proceedings of the Royal Society B* 288 (2021).

Ostrom, Elinor. *Governing the Commons.* Cambridge University Press, 1990.

Owen, Megan, et al. 'Contextual Influences on Animal Decision-Making'. *Integrative Zoology* 12 (2017).

Parise, André Geremia, et al. 'Extended Cognition in Plants: Is It Possible?' *Plant Signalling and Behaviour* 15/2 (2020).

Peña-Guzmán, David. *When Animals Dream.* Princeton University Press, 2022.

Pennis, Elizabeth. 'How Bighorn Sheep Use Crowdsourcing to Find Food on the Hoof'. *Science* (6 September 2018).

Reinhold, Annika Stefanie, et al. 'Behavioral and Neural Correlates of Hide-and-Seek in Rats'. *Science* 365 (2019).

Rößler, Daniela, et al. 'Regularly Occurring Bouts of Retinal Movements Suggest an REM Sleep-like State in Jumping Spiders'. *PNAS* 119/33 (2022).

Schluessel, V., et al. 'Cichlids and Stingrays can Add and Subtract "One" in the Number Space from One to Five'. *Nature Scientific Reports* 12 (2022).

Scott, James. *Seeing Like a State*. Yale University Press, 1998.

Shine, Rick. *Cane Toad Wars*. University of California Press, 2018.

Strandburg-Peshkin, Ariana, et al. 'Shared Decision-Making Drives Collective Movement in Wild Baboons'. *Science* 348 (2015).

Tero, Atsushi, et al. 'Rules for Biologically Inspired Adaptive Network Design'. *Science* 327 (2010).

Waal, Frans de. *Are We Smart Enough to Know How Smart Animals Are?* Granta, 2016.

Wall, Derek. *Elinor Olstrom's Rules for Radicals*. Pluto Press, 2017.

Watson, Richard, and Michael Levin. 'The Collective Intelligence of Evolution and Development'. *Collective Intelligence* 2/2 (2023).

Weintrobe, Sally. 'Moral Injury, the Culture of Uncare and the Climate Bubble'. *Journal of Social Work Practice* 34/4 (2020).

Whiten, Andrew, et al. 'The Emergence of Collective Knowledge and Cumulative Culture in Animals, Humans, and Machines'. *Philosophical Transactions of the Royal Society B* 377 (2021).

Wilson, David Sloan. *Does Altruism Exist?* Yale University Press, 2015.

Wong, Bob, and Ulrika Candolin. 'Behavioural Responses to Changing Environments'. *Behavioural Ecology* 26/3 (2015).

Yu, Bin. 'Different Neural Circuitry is Involved in Physiological and Psychological Stress-Induced PTSD-like "Nightmares" in Rats'. *Scientific Reports* 5 (2015).

Yuste, Rafael, and Michael Levin. 'New Clues About the Origins of Biological Intelligence'. *Scientific American* (11 September 2021).

LIST OF CITED SOURCES

6: Wild Clocks

Åkesson, Susanne, et al. 'Timing Avian Long-Distance Migration: From Internal Clock Mechanisms to Global Flights'. *Philosophical Transactions of the Royal Society B* 372 (2017).

Auden, W. H. *Selected Poems*. Faber, 2010.

Bartky, Ian. 'The Adoption of Standard Time'. *Technology and Culture* 30/1 (1989).

Bastian, Michelle. 'Fatally Confused: Telling the Time in the Midst of Ecological Crises'. *Environmental Philosophy* 9/1 (2012).

Bastian, Michelle. 'Liberating Clocks'. *New Formations* 92 (2017).

Bastian, Michelle, and Rowan Bayliss Hawitt. 'Multi-Species, Ecological and Climate Change Temporalities: Opening a Dialogue with Phenology'. *Environment and Planning E: Nature and Space* 6/2 (2023).

Bastian, Michelle, and Larissa Pschetz. 'Temporal Designs: Rethinking Time in Design'. *Design Studies* 56 (2018).

Benjaminsen, Tor, et al. 'Misreading the Arctic Landscape: A Political Ecology of Reindeer, Carrying Capacities, and Overstocking in Finnmark, Norway'. *Norsk Geografisk Tidsskrift* 69/4 (2015).

Betts, Jonathan. 'John Harrison: Inventor of the Precision Timekeeper'. *Endeavour* 17/4 (1993).

Brantley, Susan. 'Understanding Soil Time'. *Science* 321 (2008).

Case, Kristen. 'Knowing as Neighbouring: Approaching Thoreau's Kalendar'. *J19: The Journal of Nineteenth Century Americanists* 2/1 (2014).

Chen, I-Ching, et al. 'Rapid Range Shifts of Species Associated with High Levels of Climate Warming'. *Science* 333 (2011).

Collins, Courtney, et al. 'Experimental Warming Differentially

Affects Vegetative and Reproductive Phenology of Tundra Plants'. *Nature Communications* 12 (2012).

CAConrad. *(Soma)tic Poetry Rituals*. At: <www.somaticpoetryexercises.blogspot.com/>.

Denlinger, David, et al. 'Keeping Time Without a Spine'. *Philosophical Transactions of the Royal Society B* 372 (2017).

Emerson, Ralph Waldo. *Emerson's Prose and Poetry*. W. W. Norton & Co., 2001.

Future Library. At: <https://www.futurelibrary.no/>.

Glennie, Paul, and Nigel Thrift. *Shaping the Day: A History of Timekeeping in England and Wales 1300–1800*. Oxford University Press, 2009.

Hansen, Hanna Horsberg. 'Pile O'Sápmi and the Connection Between Art and Politics'. *SYNNYT/Origins* (2019).

Hatfield, Samantha Chisholm, et al. 'Indian Time: Time, Seasonality, and Culture in Traditional Ecological Knowledge of Climate Change'. *Ecological Processes* 7/25 (2018).

Hecksher, Christopher. 'A Nearctic-Neotropical Migratory Songbird's Nesting Phenology and Clutch Size are Predictors of Accumulated Cyclone Energy'. *Scientific Reports* (2018).

Helm, Barbara, et al. 'Annual Rhythms that Underlie Phenology: Biological Time-Keeping Meets Environmental Change'. *Proceedings of the Royal Society B* 380 (2013).

Helm, Barbara, et al. 'Two Sides of a Coin: Ecological and Chronobiological Perspectives on Timing in the Wild'. *Philosophical Transactions of the Royal Society B* 372 (2017).

Intergovernmental Panel on Climate Change, *Climate Change 2022: Impacts, Adaptations and Vulnerability*. IPCC, 2007.

Klarsfeld, Andre. 'At the Dawn of Chronobiology'. At: <http://www.bibnum.education.fr/sites/default/files/122-mairan-analysis.pdf>.

Krokene, P., et al. 'Effect of Phenology on Susceptibility of

Norway Spruce (*Picea abies*) to Fungal Pathogens'. *Plant Pathology* 61 (2012).

Kronfeld-Schor, Noga, et al. 'Chronobiology of Interspecific Interactions in a Changing World'. *Philosophical Transactions of the Royal Society B* 372 (2017).

Krznaric, Roman. *The Good Ancestor*. Penguin, 2021.

Lewin, Roger. *Making Waves*. Penguin, 2005.

Macfarlane, R. Ashton. 'Wild Laboratories of Climate Change: Plants, Phenology, and Global Warming, 1955–1980'. *Journal of the History of Biology* 54 (2021).

Mandelstam, Osip. *Journey to Armenia*. Notting Hill Editions, 2011.

Miller-Rushing, Abraham, and Richard Primack. 'Global Warming and Flower Times in Thoreau's Concord'. *Ecology* 89/2 (2008).

Nanni, Giordano. *The Colonisation of Time*. Manchester University Press, 2012.

Oskal, Nils. 'On Nature and Reindeer Luck'. *Rangifer* 20/2–3 (1999).

Our World in Data. At: <https://www.ourworldindata.org>.

Padel, Ruth. *The Mara Crossing*. Random House, 2012.

Reinert, Hugo, et al. 'The Skulls and the Dancing Pig: Notes on Apocalyptic Violence'. *Terrain* 71 (2019).

Rosemartin, Alyssa, et al. 'Lilac and Honeysuckle Phenology Data 1956–2014'. *Scientific Data* 38 (2015).

Schwartz, William, et al. 'Wild Clocks'. *Philosophical Transactions of the Royal Society B* 372 (2017).

Severson, John, et al. 'Spring Phenology Drives Range Shifts in a Migratory Arctic Ungulate with Key Implications for the Future'. *Global Change Biology* 27 (2021).

Thoreau, Henry. The Writings of Henry D. Thoreau. At: <https://thoreau.library.ucsb.edu/>.

Tyler, Nicholas, et al. 'The Shrinking Resource of Base

Pastoralism: Saami Reindeer Husbandry in a Climate of Change'. *Frontiers in Sustainable Food Systems* 4 (2021).

Walser, Robert Young. 'Dreg Songs Lost . . . and Found'. *Folk Life* 53 (2015).

Whyte, Kyle Powys. 'Time as Kinship'. In: Jeffrey Jerome Cohen, Stephanie Foote (eds), *Cambridge Companion to the Environmental Humanities*. Cambridge University Press, 2021.

Yamaguchi, Ryohei, et al. 'Trophic Levels Decoupling Drives Future Changes in Phytoplankton Bloom Phenology'. *Nature Climate Change* 12 (2022).

7: *The Lion-Man's Leap*

Anthony, Ken, et al. 'New Interventions are Needed to Save Coral Reefs'. *Nature, Ecology and Evolution* 1 (2017).

Blackiston, Douglas, et al. 'A Cellular Platform for the Development of Synthetic Living Machines'. *Science Robotics* 6 (2021).

Buchthel, Joanna, et al. 'Mice Against Ticks: An Experimental Community-Guided Effort to Prevent Tick-Borne Disease by Altering the Shared Environment'. *Philosophical Transactions of the Royal Society B* 374 (2018).

Chauhan, Nikita. 'Case Studies: Successful Wastewater Treatment through Bioremediation'. *Medium* (27 September 2023).

Crochet Coral Reef. At: <https://crochetcoralreef.org>.

Dawul Wuru website. At: <www.dawulwuru.com.au/>.

Esvelt, Kevin, and Neil Gemmell. 'Conservation Demands Safe Gene Drive'. *PLOS Biology* 15/11 (2017).

Filbee-Dexter, Karen, and Anna Smajdor. 'Ethics of Assisted Evolution in Marine Conservation'. *Frontiers in Marine Science* 6 (2019).

LIST OF CITED SOURCES

Garner, J. B. 'Genomic Selection Improves Heat Tolerance in Dairy Cattle'. *Nature Scientific Reports* 6 (2016).

Gilbert, Scott. 'A Symbiotic View of Life: We Have Never Been Individuals'. *Quarterly Review of Biology* 87/4 (2012).

Harkness, M., et al. 'In Situ Stimulation of Aerobic PCB Biodegradation in Hudson River Sediments'. *Science* 259 (1993).

Hein, Wulf. 'Tusks and Tools – Experiments in Carving Mammoth Ivory'. *L'anthropologie* 122 (2018).

Hudson, Mai, et al. 'Indigenous Perspectives and Gene Editing in Aotearoa New Zealand'. *Frontiers in Bioengineering and Biotechnology* 7/70 (2019).

ICUN. *Genetic Frontiers for Conservation*. ICUN, 2019.

Kind, Claus-Joachim, et al. 'The Smile of the Lion-Man'. *Quartär* 61 (2014).

Kofler, Natalie, et al. 'Editing Nature: Local Roots of Global Governance'. *Science* 362 (2018).

Kubis, Armin, and Arren Bar-Even. 'Synthetic Biology Approaches for Improving Photosynthesis'. *Journal of Experimental Botany* 70/5 (2019).

Malcolm, Tame. 'Is Poisoning Pests the Māori Way?' *The Spinoff* (14 March 2022).

Marris, Emma. *Wild Souls*. Bloomsbury, 2021.

Mead, Hirini Moko. *Tikanga Māori: Living by Māori Values*. Huia, 2003.

Palmer, Symon, et al. 'Gene Drive and RNAi Technologies: A Bio-cultural Review of Next-Generation Tools for Pest Wasp Management in New Zealand'. *Journal of the Royal Society of New Zealand* 532/5 (2022).

Pirsig, Wolfgang, and Kurt Wehrberger. 'The Ears of the Lion Man'. *History of Medicine Otorhinolaryngology* 1 (2015).

Preston, Christopher. 'Ethics, Experts and the Public in the Synthetic Age'. *Issues in Science and Technology* 36/3 (2020).

Quigley, Kate, et al. 'The Active Spread of Adaptive Variation for Reef Resilience'. *Ecology and Evolution* 9 (2019).

Roberts, Rome Mere. 'Walking Backwards into the Future: Māori Views on Genetically Modified Organisms'. *WINHEC* 1 (2005).

Roosth, Sophia. 'Evolutionary Yarns in Seahorse Valley'. *differences* 23/5 (2012).

Roosth, Sophia. *Synthetic: How Life Got Made.* Chicago University Press, 2017.

Ruatapu, Mohi. *Ngā Kōrero a Mohi Ruatapu.* Canterbury University Press, 1993.

Rylott, Elizabeth, and Neil Bruce. 'How Synthetic Biology Can Help Bioremediation'. *Current Opinion in Chemical Biology* 58 (2020).

Sandler, Ronald. 'The Ethics of Genetic Engineering and Gene Drives in Conservation'. *Conservation Biology* 34/2 (2019).

Shapiro, Beth. *Life as We Made It.* One World, 2021.

Singleton, Gavin. 'The Great Barrier Reef as a Cultural Landscape'. In: Pat Hutchings et al. (eds), *Coral Reefs of Australia.* CSIRO Publishing, 2022.

Spectre, Michael. 'Rewriting the Code of Life'. *New Yorker* (25 December 2016).

Torda, Gergely, et al. 'Rapid Adaptive Responses to Climate Change in Corals'. *Nature Climate Change* 7 (2017).

van Oppen, Madeleine, et al. 'Building Coral Reef Resilience Through Assisted Evolution'. *PNAS* 112/8 (2015).

van Oppen, Madeleine, et al. 'Shifting Paradigms in Restoration of the World's Coral Reefs'. *Global Change Biology* 23 (2017).

van Oppen, Madeleine, and Linda Blackall. 'Coral Microbiome Dynamics, Functions and Design in a Changing World'. *Nature Reviews Microbiology* 17 (2019).

Wright, Robyn, et al. 'A Multi-OMIC Characterisation of

LIST OF CITED SOURCES

Biodegradation and Microbial Community Succession with the PET Plastisphere'. *Microbiome* 9 (2021).

Yoshida, Shosuke, et al. 'A Bacterium that Degrades and Assimilates Poly(ethylene terephthalate)'. *Science* 351 (2016).

Zrimec, Jan, et al. 'Plastic-Degrading Potential Across the Global Microbiome Correlates with Recent Pollution Trends'. *mBio* 12 (2012).

Kafka's Leopards

Kafka, Franz. *The Blue Octavo Notebooks*. Exact Change, 1991.

IMAGE CREDITS

p.x Cliff swallow with shorter, blunter wings. © Glenn Bartley Nature Photography

p.12 Mechta ('Dream'), the first silver fox to develop floppy ears. © Dugatkin, L.A. The silver fox domestication experiment. Evo Edu Outreach 11, 16 (2018). https://doi.org/10.1186/s12052-018-0090-x Reprinted by permission of the Creative Commons Attribution 4.0 International Licence: https://creativecommons.org/licenses/by/4.0/

p.40 Magpie nest made out of 1,500 metal spikes. © Auke-Florian Hiemstra

p.74 Aguahoja III (detail). Reprinted by permission of Kelly Egorova at Oxman.

p.106 Megan Watts Hughes, three-pitch Impression Figure ('Octave and 5th interval B flat') made with an Eidophone. © Louis Porter & Cyfarthfa Castle Museum and Art Gallery

p.142 The 'Barracuda Effect'. © Wikimedia Commons

p.178 The Silent Room at the Future Library. © Katie Paterson, Future Library, 2014-2114. Silent Room. Photo by Einar Aslaksen, 2023

p.214 The Lion-Man of Hohlenstein-Stadel, Lonetal (Bade-Wurtemberg). © Dagmar Hollmann / WikimediaCommons. License: CC BY-SA 4.0 (https://creativecommons.org/licenses/by-sa/4.0/legalcode)

ACKNOWLEDGEMENTS

Like Claude Lévi Strauss, I believe that 'animals are good to think with.' But so, in many different ways, are people. My thanks to:

The brilliant team at Canongate, especially Simon Thorogood, Francis Bickmore, Claire Reiderman, Jenny Fry, Amaani Banharally, Caitriona Horne, Vicki Rutherford and Sylvie the Canongate dog.

Gemma Wain, for copyediting excellence.

Stephen Parker, for such a striking and eloquent cover (a frankincense tree, Boswellia sacra, clinging tenaciously to a precipice on Socotra Island, Yemen).

My wonderful agent Carrie Plitt.

Richard Fisher, for commissioning the essay for BBC Future from which this book grew.

For kindness in Oslo: Anne Beate Hovind and Katie Paterson.

For kindness in New Jersey, New York and Rhode Island: Suzanne Rabb Green, Rebecca Altman, and Karen Bishop.

For their time and expertise: Parvez Alam, Leonie Alexander, Gesine Argent, Rachel Armstrong, Duncan Baker-Brown, Michelle Bastian, Janine Benyus, Doug Blackiston, Aaron Bradshaw, Jane Calvert, Louisa Casson, Holly Corfield Carr, Adam Dickinson, Ziv Epstein, Kevin Esvelt, Simone Ferracina, Eric Fisher, Jessica Gaitán Johannesson, Duncan Geere, Lily

Green, Siddhartha Hayes, Barbara Helm, Lisa Houston, Cora Kreikamp, Nic Lee, Michael Levin, Tame Malcolm, Melanie Mark-Shadbolt, George Monbiot, Shannon Nangle, Stephen Palumbi, Michael Pawlyn, Miriam Quick, Carrie Roble, Menno Schilthuizen, Marcus Shadbolt, Dav Shand, Beth Shapiro, Gavin Singleton, Erika Szymanski, Chris Thomas, Peter Tyack, Madeleine van Oppen, Ane Victoria Vollsnes, Tim Vincent-Smith (and Hannah Kitchen-Kirby for the connection), Margaret Wertheim, Andrew Whitehead, Ike Wirgin, and En Ze Linda Zhong-Johnson.

Most of all, Rachel, Isaac and Annie – you are my life's genius.